FM 3-05.230

Special Forces Tactical Facilities

February 2009

DISTRIBUTION RESTRICTION: Distribution authorized to U.S. Government agencies and their contractors only to protect technical or operational information from automatic dissemination under the International Exchange Program or by other means. This determination was made on 14 November 2008. Other requests for this document must be referred to Commander, United States Army John F. Kennedy Special Warfare Center and School, ATTN: AOJK-DTD-SF, Fort Bragg, NC 28310-9610, or higher authority.

DESTRUCTION NOTICE: Destroy by any method that will prevent disclosure of contents or reconstruction of the document.

FOREIGN DISCLOSURE RESTRICTION (FD 6): This publication has been reviewed by the publication developers in coordination with the United States Army John F. Kennedy Special Warfare Center and School foreign disclosure authority. This publication is releasable to students from foreign countries on a case-by-case basis only.

Headquarters, Department of the Army

This publication is available at
Army Knowledge Online (www.us.army.mil) and
General Dennis J. Reimer Training and Doctrine
Digital Library at (www.train.army.mil).

*FM 3-05.230

Field Manual
No. 3-05.230

Headquarters
Department of the Army
Washington, DC, 8 February 2009

Special Forces Tactical Facilities

Contents

Page

PREFACE v

Chapter 1	THE SPECIAL FORCES TACTICAL FACILITY 1-1
	Overview... 1-1
	Three Phases of Special Forces Tactical Facility Development..................... 1-2
	Tactical Facility Essential Support Systems 1-2
	Tactical Facility Critical Nodes Matrix 1-3
	Rural Tactical Facility.. 1-4
	Urban Tactical Facility... 1-12
	Summary 1-13
Chapter 2	TACTICAL FACILITY PLANNING AND DESIGN... 2-1
	Section I—Planning.. 2-1
	Statement of Requirements... 2-1
	Area Study.. 2-1
	Site Survey 2-1
	Area Assessment 2-2
	Special Operations Debrief and Retrieval System.. 2-2
	Section II—Design.. 2-3
	Inner Perimeter... 2-3
	Inner Barrier and Outer Barrier.. 2-12
	Outer Perimeter.. 2-14
	Administration Area 2-19

Distribution Restriction: Distribution authorized to U.S. Government agencies and their contractors only to protect technical or operational information from automatic dissemination under the International Exchange Program or by other means. This determination was made on 14 November 2008. Other requests for this document must be referred to Commander, United States Army John F. Kennedy Special Warfare Center and School, ATTN: AOJK-DTD-SF, Fort Bragg, NC 28310-9610, or higher authority.

Destruction Notice: Destroy by any method that will prevent disclosure of contents or reconstruction of the document.

Foreign Disclosure Restriction (FD 6): This publication has been reviewed by the publication developers in coordination with the United States Army John F. Kennedy Special Warfare Center and School foreign disclosure authority. This publication is releasable to students from foreign countries on a case-by-case basis only.

*This publication supersedes FM 3-05.230, 30 July 2003.

Contents

	Access Road	2-24
	Surrounding Area	2-24
Chapter 3	**TACTICAL FACILITY CONSTRUCTION**	**3-1**
	Construction Principles	3-1
	General Structural Components	3-1
	Concrete	3-8
	Rock	3-8
	Brick and Masonry	3-9
	Wood	3-9
	Sandbags	3-9
	Pallets	3-10
	Ammunition Containers	3-10
	Steel Drums	3-10
	Large Shipping Containers	3-11
	Metal Pipe	3-11
	Cinder Blocks	3-11
	Fencing	3-12
	Emergency Signal Project	3-12
Chapter 4	**TACTICAL FACILITY OPERATIONS AND DEFENSE**	**4-1**
	Areas of Responsibility	4-1
	Operational Environments	4-1
	Threat Level	4-2
	Joint Fires	4-3
	Indirect Fire	4-4
	Close Combat Attack	4-4
	Close Air Support	4-5
	Radars	4-5
	Unmanned Aircraft Systems	4-9
	Multinational Fire Support	4-9
	Fire Support Planning	4-10
Chapter 5	**TACTICAL FACILITY SUSTAINMENT**	**5-1**
	Logistics Overview	5-1
	Developed Theater Logistics	5-1
	Undeveloped Theater Logistics	5-5
	Special Forces Operational Detachment Support and Sustainment	5-7
Chapter 6	**TACTICAL FACILITY FUNDING**	**6-1**
	Funding Principles	6-1
	Contracting Process	6-2
Chapter 7	**TACTICAL FACILITY RELIEF IN PLACE**	**7-1**
	Procedures	7-1
	Considerations	7-1
	Types of Relief in Place or Closeout	7-3
	Relief in Place Environmental Factors	7-4
	Closeout Assistance	7-6
	Final U.S. Closeout	7-7

Contents

Appendix A	AREA STUDY	A-1
Appendix B	SITE SURVEY	B-1
Appendix C	AREA ASSESSMENT	C-1
	GLOSSARY	Glossary-1
	REFERENCES	References-1
	INDEX	Index-1

Figures

Figure 1-1. Special Forces tactical facility critical nodes matrix 1-3
Figure 1-2. Rural Special Forces tactical facility critical nodes 1-5
Figure 1-3. Initial phase rural tactical facility (Afghanistan) 1-6
Figure 1-4. Initial phase rural tactical facility rudimentary housing (Africa) 1-6
Figure 1-5. Initial phase rural tactical facility shower facility 1-7
Figure 1-6. Initial phase rural tactical facility latrine 1-7
Figure 1-7. Temporary phase rural tactical facility wood-framed structures 1-8
Figure 1-8. Temporary phase latrines 1-9
Figure 1-9. Power transformers routing electricity to structures 1-9
Figure 1-10. Redundant permanent phase generators 1-10
Figure 1-11. Filling stackable barrier systems 1-11
Figure 1-12. Tactical facility improved protection measure—solid, defensible mud wall 1-11
Figure 1-13. Urban Special Forces tactical facility critical nodes 1-12
Figure 2-1. Rural Special Forces tactical facility overview (example 1) 2-4
Figure 2-2. Rural Special Forces tactical facility overview (example 2) 2-4
Figure 2-3. Inner protective berm 2-5
Figure 2-4. Elevated inner berm positions 2-5
Figure 2-5. Rocket predetonation sequence 2-6
Figure 2-6. DA Form 5517-R (Standard Range Card) 2-8
Figure 2-7. Personnel bunker offering indirect-fire protection 2-9
Figure 2-8. Mortar position 2-11
Figure 2-9. Machine gun bunker 2-15
Figure 2-10. Sandbags stacked chest-high around tactical facility structures 2-16
Figure 2-11. Outer perimeter corner observation tower 2-17
Figure 2-12. Fuel point 2-19
Figure 2-13. Helicopters landing at a tactical facility helicopter landing zone 2-20
Figure 2-14. Host-nation dispensary 2-20
Figure 2-15. Constructing a rebar reinforced concrete helicopter pad 2-21
Figure 2-16. Vetted host-nation employees wearing colored vests 2-22
Figure 2-17. Main gate entrance and confinement zone 2-23

Figure 2-18. Access road 2-24
Figure 3-1. Mortar pit with mil markings 3-3
Figure 3-2. Mortar pit without mil markings 3-3
Figure 3-3. Observation tower 3-4
Figure 3-4. Indigenous employees mixing cement 3-5
Figure 3-5. Brick slab roofing 3-9
Figure 4-1. Close combat attack checklist 4-5
Figure 4-2. Unattended transient acoustic measurement and signature
   intelligence system 4-6
Figure 4-3. Lightweight countermortar radar 4-7
Figure 4-4. AN/TPQ-36 radar 4-7
Figure 4-5. AN/TPQ-37 radar 4-8
Figure 4-6. MQ-1 Predator unmanned aircraft system 4-9
Figure 4-7. Standard 9-line report format 4-11
Figure 4-8. Special Forces immediate close air support request channels 4-12
Figure 4-9. Fixed-wing airspace control measures 4-17
Figure 4-10. Formal airspace coordination area 4-18
Figure 5-1. Airdrop resupply 5-3
Figure 5-2. Vehicle recovery 5-4
Figure 5-3. Statement of requirements format 5-9
Figure 6-1. Fiscal questionnaire 6-6
Figure 6-2. Resourcing funds and contracting checklist 6-14
Figure A-1. Special Forces area study A-1
Figure B-1. Site survey checklist B-1
Figure C-1. Initial area assessment C-1
Figure C-2. Principal area assessment C-2
Figure C-3. Preventative medicine area assessment C-5

Tables

Table 3-1. Minimum thickness (in inches) required for protection against
   enemy fire 3-6
Table 3-2. Minimum thickness (in inches) required for protection against
   high-explosive shaped charges 3-7

Preface

Field Manual (FM) 3-05.230, *Special Forces Tactical Facilities*, supports key United States (U.S.) Army Special Forces (SF) doctrine. An SF tactical facility (TACFAC) is defined as any secure urban or rural facility that enables Army special operations forces (ARSOF) to extend command and control (C2), provides support for operations, and allows operational elements to influence a specified area. SF TACFACs include a variety of secure locations for SF operations, including (but not limited to) firebases, camps, and team houses.

PURPOSE

As with all doctrinal manuals, FM 3-05.230 is authoritative but not directive. It serves as a guide but does not preclude SF units from developing their own standing operating procedures (SOPs) to meet their needs. This FM focuses on the establishment, improvement, operations, and security of SF TACFACs.

SCOPE

This FM presents the details of the three phases of SF TACFAC development in an order that the SF Soldier should expect to encounter them—initial, temporary, and permanent. Additionally, this FM discusses in sequence planning, design, construction, operations, sustainment, funding, and transfer of authority. The primary audiences of this FM are the commanders, staff officers, and operational personnel of Special Forces operational detachments A (SFODAs), Special Forces operational detachments B (SFODBs), and Special Forces operational detachments C (SFODCs).

APPLICABILITY

Commanders and trainers should use this and other related manuals in conjunction with command guidance and the mission training plan (MTP) to plan, construct, and operate from SF TACFACs. This FM is applicable to the Active Army, the Army National Guard (ARNG)/Army National Guard of the United States (ARNGUS), and the United States Army Reserve (USAR) unless otherwise stated.

ADMINISTRATIVE INFORMATION

The proponent of this FM is the United States Army John F. Kennedy Special Warfare Center and School (USAJFKSWCS). Submit comments and recommended changes to Commander, USAJFKSWCS, ATTN: AOJK-DTD-SF, Fort Bragg, NC 28310-9610. This FM is designed to be UNCLASSIFIED in order to ensure the widest distribution possible to the appropriate ARSOF and other interested Department of Defense (DOD) and United States Government (USG) agencies while protecting technical or operational information from automatic dissemination under the International Exchange Program or by other means. Unless this publication states otherwise, masculine nouns and pronouns do not refer exclusively to men.

This page intentionally left blank.

Chapter 1

The Special Forces Tactical Facility

The primary role of the SF TACFAC is to support special operations (SO) and function as a tactical and operational base. As such, TACFACs serve a primarily defensive function, although they may serve as bases for needed offensive operations. The secondary role of the TACFAC is to be a center that develops, nurtures, and maintains liaison with local host-nation (HN) populace and members of the HN military and civilian leadership. This role is critical in foreign internal defense (FID) and counterinsurgency (COIN) operations.

Over time, the TACFAC helps provide for establishment, restoration, and improvements of many local HN community and governmental services and systems. These essential support systems for the TACFAC and surrounding HN communities are best captured by the acronym SWEAT-MSS (security, water, electricity, administration, trash, medical, sewage, and shelter). Eventually, the SF TACFAC will be returned to the control of the HN government through a relief in place (RIP).

OVERVIEW

1-1. SF operations support the operations and goals of the geographic combatant commanders (GCCs) and their subordinate joint force commanders (JFCs). SF TACFACs primarily serve as a base to support SF operations. As such, they serve an operationally defensive role. They allow SF units and partnered HN or multinational forces to meet the tasks and purposes of the defense and aid in setting the conditions for successful offensive operations, which can be either lethal or nonlethal. FM 3-0, *Operations*, defines defensive operations as "combat operations conducted to defeat an enemy attack, gain time, economize forces, and develop favorable conditions for offensive or stability operations."

1-2. Several of the purposes of defensive operations are very closely related to SF TACFAC operations. These purposes include the following:
- Deter and defeat enemy attacks (such as insurgency operations).
- Achieve economy of force (such as allowing the JFC to concentrate conventional forces elsewhere against an enemy main effort).
- Retain key terrain (which includes the local populace).
- Protect the populace, critical assets, and infrastructure.
- Develop intelligence (particularly local intelligence).

1-3. SF TACFAC operations also can have a tremendous positive impact in support of friendly information operations (IO), as TACFACs are tangible proof of a stabilizing friendly-force presence. In order for U.S. Armed Forces—and SF in particular—to have any positive effect on the local population's view of their government, local critical assets and infrastructure must be defended. These assets and infrastructure usually have more economic and political value than tactical military value.

1-4. Prior to selection of the TACFAC site, the Special Forces operational detachment (SFOD) must analyze the mission, enemy, terrain and weather, troops and support available, time available, and civil considerations (METT-TC). When planning and designing the TACFAC, the SFOD must take into account the factors critical to the security of the facility, including observation and fields of fire, avenues of approach, key terrain, obstacles, and cover and concealment (OAKOC). The SF TACFAC commander uses all available assets during defensive planning. These include intelligence, reconnaissance, and engineer assets to study the terrain. Optimum use of terrain, depth, and security operations allows the SFOD to minimize defensive resources.

1-5. The smallest SF unit tasked to operate and maintain an SF TACFAC is the SFODA. SFODBs and SFODCs that occupy TACFACs produce much larger tactical footprints in their areas of operation (AOs), which may be a distinct disadvantage. The larger advanced operating base (AOB) or special operations task force (SOTF) TACFAC needed to sustain a greater force requires additional logistics and protection support.

1-6. SF units must determine if there are advantages or disadvantages to breaking ground on a new facility or using an existing facility. Questions that are considered when making this decision include the following:

- Is the AO permissive, uncertain, or hostile?
- Are the facilities located in urban or rural settings?
- Are the facilities logistically sustainable?
- Are HN defense, security, and protection adequate?
- Are there established casualty and emergency evacuation plans?
- Are there future plans to close, convert, or abandon the facility?
- Are the facilities part of a larger operation with a yet-unspecified strategic plan?
- Are there Civil Affairs (CA) and Psychological Operations (PSYOP) missions in the AO?

1-7. In addition to the SFOD (or SFODs) occupying the TACFAC, the facility may also house vetted interpreters and other friendly force personnel on either a permanent or temporary basis. These friendly forces may include other U.S. or multinational force personnel (military, interagency, or contractor), HN personnel (military, constabulary, or civilian government), or irregular forces working with the SFOD. The SFOD must take into account both the size of these elements and any cultural considerations related to these various groups when developing a TACFAC.

THREE PHASES OF SPECIAL FORCES TACTICAL FACILITY DEVELOPMENT

1-8. The three broad phases of TACFAC development are initial, temporary, and permanent, whether in a rural or urban environment. Therefore, SF TACFACs are classified by phase and by environment, such as initial (rural) and initial (urban). The transformation of the TACFAC through these phases is a mission-dependent process, and it is possible to begin a TACFAC at a level higher than the initial (bare base) phase. For example, an SFOD may rent, lease, or otherwise occupy a compound that was already developed into a defensible, habitable place. In such cases, some of the basic support systems (described in paragraph 1-10, below) are already in place and functioning. Therefore, development of a specific TACFAC may begin at the temporary (intermediate) or even permanent (well-developed) phase. Generally, this is more common in urban TACFACs than in rural TACFACs, as urban areas contain numerous existing structures and are more likely to have some forms of required supporting infrastructure.

1-9. SF units operate in a wide range of environmental conditions—from the deserts and cities of Iraq to the mountains and valleys of Afghanistan to the jungles of Colombia and the Philippines. Because of this, SF TACFACs may be further classified by specific environmental conditions, such as rural (desert) or rural (jungle). Although a TACFAC located in Iraq may appear quite different from a TACFAC in the Philippines, both sites essentially function alike (supporting SF operations) and have similar support systems requirements.

TACTICAL FACILITY ESSENTIAL SUPPORT SYSTEMS

1-10. The essential support systems for the TACFAC are best captured by the acronym SWEAT-MSS. These support systems also are of critical importance to the surrounding HN communities, and improving the essential support systems of the TACFAC can provide the basis for improving those systems for the local populace. This point is best illustrated by FM 3-07, *Stability Operations and Support Operations*.

which uses the acronym SWEAT-MS (sewage, water, electricity, academics, trash, medical, and security) to describe U.S. Army lines of effort in helping the local populations.

1-11. Although most of the SF TACFAC essential support systems are self-explanatory, there are two important points to note. Security is listed first in SWEAT-MSS because it is paramount in all tactical operations and therefore must be analyzed before anything else. Second, SWEAT-MSS uses the category of administration and HN training as an essential support system.

1-12. Administration is a relatively broad category which encompasses both the daily administrative and logistical concerns of running the TACFAC (Chapters 5 and 6 further discuss TACFAC support and funding) as well as the fundamental role that SF operations—based out of the TACFAC—play in enhancing the HN government role in developing and administering political, economic, and security control in the local area. As such, the SFOD must plan for HN training requirements—tactical, administrative, and logistical. This includes having areas to meet with members of the HN populace—even an initial phase TACFAC requires administrative areas to interact with HN government officials, tribal leaders, and local contractors. These areas are located inside the outer perimeter.

TACTICAL FACILITY CRITICAL NODES MATRIX

1-13. The SF TACFAC critical nodes matrix (CNM) (Figure 1-1) is a planning guide used to assist an SF element by providing a starting point to not only establish an SF TACFAC, but to modify or improve an existing facility. The TACFAC CNM identifies the environment and different developmental phases. The environment is defined as either rural or urban, and the classifications or phases of development are described as initial, temporary, or permanent.

	Initial	Temporary	Permanent
Security and Protection			
Water			
Electricity			
Administration and Host-Nation Training			
Trash			
Medical			
Sewage			
Shelter			

Figure 1-1. Special Forces tactical facility critical nodes matrix

1-14. The TACFAC CNM analyzes specific requirements across the TACFAC essential support systems to identify critical requirements through the phases of TACFAC development. Much information can be drawn from the CNM through analysis of critical nodes or elements required to sustain or maintain a TACFAC and its critical systems. Through this process a logical progression is established between the different phases which helps guide commanders and staffs at all levels in planning for and supporting the development of the TACFAC. This is a major advantage for SF units, particularly during more intense and fluid operations, such as active COIN environments.

1-15. SFODs normally require additional nonorganic support to properly run their TACFACs and to allow them to properly function at full operational capacity. Without this support, the JFC will not get the full benefit of the SFOD's very significant operational effects. The CNM helps in identifying existing shortfalls or gaps in personnel, equipment, and materials needed in order to establish or modify an existing TACFAC. By identifying these shortfalls or gaps early during the analysis process they can be properly forecasted or requested. The CNM can help to identify such potential personnel shortfalls as vehicle mechanics, generator mechanics, cooks, and medical augmentation.

1-16. As with personnel analysis, the CNM can help identify critical operational shortfalls. For example, a security systems analysis may identify the need for such tactical force augmentation as howitzer sections, counterfire radar systems, thermal imaging systems, or even infantry or military police forces. An administration systems analysis may identify requirements and opportunities for a PSYOP or CA team. The CNM not only benefits the end user, but is a useful tool for the supporting staff to identify, plan, and forecast future items needed to support the end users.

1-17. The second advantage to using the CNM is that it allows the commander to identify and analyze critical nodes or resources required to maintain or sustain an SF TACFAC and its critical support systems (SWEAT-MSS). Using this analysis process, a logical matrix support planning progression is established with primary, alternate, contingency, emergency (PACE) plans.

1-18. If an existing critical node in the SWEAT-MSS model fails anywhere within the matrix, the previous critical node is used as a substitute. For example, a rural SF TACFAC is in the final permanent development phase and the sewage plan fails. According to the PACE concept and the CNM, the alternate solution is to use the burn-barrel node of the previous (temporary) TACFAC development phase. If the alternate plan fails, the detachment uses the slit-trench node from the initial TACFAC development phase. This cascade of options is logical and controlled, and it serves the detachment until such time as the primary critical node system is repaired or replaced.

RURAL TACTICAL FACILITY

1-19. The rural TACFAC may progress through all three phases of development in order, it may remain at one phase, or it may skip phases. When a TACFAC is first occupied by an arriving SF unit, it becomes—by default—the initial TACFAC. Critical nodes for a rural TACFAC development are described in the example in Figure 1-2, page 1-5.

INITIAL PHASE

1-20. The initial development phase of a rural TACFAC begins when the first SF unit occupies the site. The initial facility may be used or occupied for only a short time and then abandoned or it may transition through one, two, or all three phases of development. When the initial phase begins, the facility usually is primitive and may include a large portion of undeveloped land (Figure 1-3, page 1-6). The major concerns of the SFODA during the initial development of a TACFAC include basic survival needs (for example, security, water, food, sanitation, and electric power).

1-21. Housing is rudimentary—typically in general purpose (GP) medium tents (Figure 1-4, page 1-6) or bivouac system modular tents (B-huts)—and basic protective measures are implemented (such as 24-hour guards and short-duration patrols). Electric power is supplied by small, portable generators, such as the commercially procured 5-kilowatt generators assigned to most SFODs. Solar power may be used as an initial-phase energy source, along with the SFOD's solar-powered battery rechargers. Basic comforts, such as indoor plumbing and running water, typically are nonexistent. To address water needs, a natural fresh-water source (such as a stream) should be in close proximity. Rain water and potable water is stored in closed containers to prevent potential contamination and disease, and the SFOD constructs simple showers (Figure 1-5, page 1-7) for personal hygiene. Slit trenches and cat holes are used until the SFOD can construct burn-barrel latrines (Figure 1-6, page 1-7).

	Initial	Temporary	Permanent
Security and Protection	• 24/7 SFODA security • Triple-strand concertina • Fighting positions	• Stackable barrier system walls • Sandbag bunkers • SFODA with HN augmentation	• Brick and mortar wall • Observation tower • HN augmented with SFODA
Water	• Bottled water • Water purification kits	• Underground well • ROWPU	• Water tower • Plumbing
Electricity	• Batteries • 5 kW generator	• 20 kW generator	• 200 kW generator
Administration and Host-Nation Training	• GP medium tents • Sand tables • Tape drill area	• B-hut classrooms (HN construction) • HN meeting area	• Fixed ranges • Rehearsal area
Trash	• Local burn pit	• Local disposal (operational funds)	• Contract services • Incinerator
Medical	• MOS 18D (M-5 medical bag) • Medical bunker • HLZ	• SFODA dispensary • HN treatment area	• Clinic with U.S. and HN medical augmentation
Sewage	• Slit trench	• Burn barrels	• PVC sewerage (local leach fields)
Shelter	• GP medium tents	• B-huts	• Permanent hardened structures
Legend: 18D Special Forces medical sergeant HLZ helicopter landing zone kW kilowatt MOS military occupational specialty PVC polyvinyl chloride ROWPU reverse-osmosis water purification unit			

Figure 1-2. Rural Special Forces tactical facility critical nodes

1-22. As the TACFAC develops, security is enhanced and greater protective measures are put into practice. These measures include longer patrols, listening posts (LPs), observation posts (OPs), and additional wire around the facility perimeter. After basic security is established initial construction projects—such as inner and outer perimeter barriers—are built. Early projects may be hampered by the limited equipment and material assets carried in by the SF unit. The early initial projects are planned and coordinated prior to the SFODA arrival, which facilitates TACFAC construction.

Chapter 1

Figure 1-3. Initial phase rural tactical facility (Afghanistan)

Figure 1-4. Initial phase rural tactical facility rudimentary housing (Africa)

The Special Forces Tactical Facility

Figure 1-5. Initial phase rural tactical facility shower facility

Figure 1-6. Initial phase rural tactical facility latrine

Chapter 1

TEMPORARY PHASE

1-23. As the initial developmental phase of a rural TACFAC progresses, and new buildings are constructed and infrastructure renovations are made, the initial phase transforms into the temporary phase. During this phase, wood-framed buildings (Figure 1-7) or buildings made from common local construction materials replace tents and crude huts. In a temporary TACFAC, the SF unit's standard of living improves greatly over that of the initial facility, to include such "creature comforts" as improved wooden outhouses, potable running water, air conditioners, and heaters. Typical structures built during the temporary phase include an operations center (OPCEN), medical center, dining facility (DFAC), latrine (Figure 1-8, page 1-9), shower, and workshop accommodations. In a temporary TACFAC, most buildings are wired for electricity (Figure 1-9, page 1-9). The small, portable 5-kilowatt generator is replaced by larger generators, such as a U.S. Army–issue 20-kilowatt generator. The 5-kilowatt generator then becomes the alternate backup generator.

Figure 1-7. Temporary phase rural tactical facility wood-framed structures

1-24. In a temporary phase TACFAC, the SFOD applies enhanced protection measures and expands facility control outward to the access road and surrounding areas. Security patrols go further out into the surrounding area, and the SFOD continues to increase contact with the local populace. This contact improves opportunities to engage local residents on a variety of issues, and presents numerous chances to increase support for friendly-force operations and for the HN government. During this phase, it is critical that the local populace understands that the SFOD and HN presence will bring enhanced security and the possibility for multiple development projects. As the SFOD continues to gain HN trust and support, the construction of an HN meeting area or building in the administration area of the TACFAC is well under way. The physical security of the TACFAC remains a major concern, and the SFOD must continue to improve physical security measures. Earthen berms form barriers and, if possible, stackable barrier system walls are built. As new additions and upgrades to existing infrastructure take place, the site begins to transform from a temporary to a permanent TACFAC.

The Special Forces Tactical Facility

Figure 1-8. Temporary phase latrines

Figure 1-9. Power transformers routing electricity to structures

Chapter 1

PERMANENT PHASE

1-25. The third and final phase of a rural TACFAC is the permanent phase. This phase is initiated when the SF unit expects to occupy the facility for an extended period of time. The permanent rural TACFAC is better developed, organized, maintained, and prepared to handle security emergencies than the initial or temporary TACFAC. The outer barrier may be a mud or brick wall, and security patrols are conducted at random.

1-26. The permanent rural TACFAC should have a minimum of two covered and protected 200-kilowatt diesel-electric generators (Figure 1-10). They should be of the same brand, type, and electrical capacity to minimize spare parts and optimize efficiency. If necessary, parts can be interchanged. The generator designated as primary is used for all daily electrical requirements. All other generators can provide electrical backup during scheduled outages, routine maintenance, repairs, and emergencies.

Figure 1-10. Redundant permanent phase generators

1-27. The permanent TACFAC differs from initial and temporary facilities in that the facility occupies not only one or two buildings, but an enclosed building complex. This complex offers greater security and improved protection options. Permanent facilities incorporate PACE security planning for water, electricity, medical treatment, communications, personnel security, and secure storage of ammunition and fuel. Critical spare parts should be kept on hand.

1-28. The SFODA achieves control of the access road, administration sector, and surrounding area using perimeter and barrier defenses in depth. An observation tower within the inner perimeter is a valuable asset. Multiple rows and strands of barbed wire, tanglefoot, and claymore mines should be emplaced. Stackable barrier systems (Figure 1-11, page 1-11) or fort-like barrier walls provide a great deal of security for the facility. Outer barrier walls common in the Middle East and Central Asia (Figure 1-12, page 1-11) are made from a local brown mud-type material and may be 4 to 5 meters high and 1 meter thick.

The Special Forces Tactical Facility

Figure 1-11. Filling stackable barrier systems

Figure 1-12. Tactical facility improved protection measure—solid, defensible mud wall

URBAN TACTICAL FACILITY

1-29. Urban TACFACs almost always are based on preexisting urban structures. There are unique advantages and challenges in developing an urban TACFAC. For example, an SF unit occupying an urban TACFAC may be able to take advantage of preexisting water, sewer, and electrical systems (if they are still functional). However, the urban environment does offer some significant challenges to security because of the proximity to surrounding buildings and the dense population concentration (which will likely include hostile elements). Specific threats also increase in urban environments, such as the threat of snipers and explosive devices (such as vehicleborne improvised explosive devices [VBIEDs]). Specific security measures and equipment are covered in detail in Chapter 4.

1-30. Urban TACFACs, like their rural TACFAC counterparts, also may evolve through initial, temporary, and permanent phases. Protection and security continue to remain paramount (as does the overall defensive posture of the TACFAC) by maintaining the PACE planning process in an urban environment. The critical nodes for urban TACFAC development are described in Figure 1-13.

	Initial	Temporary	Permanent
Security and Protection	• 24/7 SFODA security • Triple-strand concertina • Fighting positions	• Perimeter fence • Rooftop security • SFODA with HN augmentation	• Brick and mortar wall • Observation tower • HN augmented with SFODA
Water	• Bottled water • Water purification kits	• Local pumps • ROWPU	• Indoor plumbing
Electricity	• Batteries • 5 kW generator	• 20 kW generator	• 200 kW generator • Local grid
Administration and Host-Nation Training	• Sand tables	• Temporary classrooms • HN meeting area	• 25-meter range • Rehearsal area
Trash	• Burn pit	• Local disposal (operational funds)	• Sanitary and secure landfill (contracted) • Incinerator
Medical	• MOS 18D (M-5 medical bag) • Medical bunker • HLZ	• SFODA dispensary • HN treatment area	• Clinic with U.S. and HN medical augmentation
Sewage	• Portable toilets	• Septic tank	• Local sewage treatment plant
Shelter	• Existing structure (abandoned or leased)	• Improved structures	• Hardened structures (masonry or concrete)

Figure 1-13. Urban Special Forces tactical facility critical nodes

INITIAL PHASE

1-31. The biggest difference between the initial rural TACFAC and the initial urban TACFAC is the structure. In an urban TACFAC, the building likely already exists. Selection criteria include the following:
- Does the structure have a barrier around the perimeter, such as a fence or wall?
- Is the structure connected to the existing power grid and water and sewer systems?

1-32. Ideally, the initial structure will have most of the upgrades required to transition from an initial to a permanent TACFAC. If not, as a minimum, initial requirements include the following:
- 5-kilowatt generator.
- Indoor toilet, outhouse, or burn barrel.
- Perimeter protection (fence or brick wall). If no perimeter protection exists, continuous security patrols and observation are required.

TEMPORARY PHASE

1-33. If a perimeter wall or fence does not exist, it is installed during this phase of development. Also, the unit constructs a safe room. These additions create extra protection and explosive standoff distance. Other temporary urban upgrades include 2 or more 20-kilowatt generators, portable toilets, and extra rooftop security.

PERMANENT PHASE

1-34. In the permanent urban TACFAC phase, power-generation requirements may increase, thereby requiring a pair of 200-kilowatt generators of the same make, model, and capacity. Indoor plumbing and sewerage services are made available. Outside security is enhanced with the addition of a rooftop guard tower; permanent, industrial grade security cameras; motion-activated lights; sensors, including electro-optical devices; and increased active and passive security and surveillance measures.

SUMMARY

1-35. As the TACFAC progresses through the three phases of development in both rural and urban environments, the previous phase remains a part of the PACE plan for TACFAC operations. For example, small, portable, 5-kilowatt electric generators that were used as a primary source of power in the initial phase revert to contingency or emergency use in the temporary phase.

1-36. There are no clearly defined checklists or timelines for TACFAC development. New building construction may begin at any time during any phase, and upgrades, repairs, and maintenance should be constant. Defensive improvements and facility infrastructure improvements must be planned and developed according to METT-TC.

This page intentionally left blank.

Chapter 2
Tactical Facility Planning and Design

Commanders of SFODAs, SFODBs in AOBs, or SFODCs in SOTFs may be tasked with planning, designing, constructing, and operating from TACFACs. Each SOTF commander normally organizes his base into an OPCEN, signals center (SIGCEN), and support center (SPTCEN). Isolation is a technique, not a building; as a result, all SFOD TACFACs must be able to combine secure planning with self-isolation capabilities within their OPCEN.

SECTION I—PLANNING

2-1. Armed with the basic requirements and phases of identifying and/or constructing a TACFAC, an SF unit conducts the standard military decision-making process (MDMP). The MDMP steps found in FM 5-0, *Army Planning and Orders Production*, and Graphic Training Aid (GTA) 31-01-003, *Detachment Mission Planning Guide*, help to determine the course of action (COA) required for each specific TACFAC to become operational.

2-2. The statement of requirement (SOR), area study (Appendix A), site survey (Appendix B), area assessment (Appendix C), and special operations debrief and retrieval system (SODARS) are five of many resource documents used during the predeployment planning process. Depending upon the mission, these documents may carry a security classification of CONFIDENTIAL or higher.

STATEMENT OF REQUIREMENTS

2-3. The intent of the SOR is to identify the requesting unit's logistic and materiel needs—particularly those that exceed their organic capabilities—early in the MDMP. This early identification becomes particularly relevant when an SF unit is tasked to develop an original TACFAC.

2-4. Once developed, consolidated, and prioritized, the SF unit routes the SOR through the chain of command. This ensures that mission supply (Classes I through X), facility maintenance, transportation, personnel, medical, signal, operational, and security needs are addressed and any shortcomings are identified as the SF unit continues to plan.

AREA STUDY

2-5. The area study provides initial information to the SFOD about the specific country where they will perform operations. It is a valuable planning resource, along with the site survey, area assessment, and the SODARS. (Appendix A provides the format for an area study.)

SITE SURVEY

2-6. The site survey team operates in a similar fashion to a reconnaissance patrol. Teams typically deploy in small units of two to six Soldiers and gather information. Soldiers on the survey team are from the unit that will occupy the selected site. The team conducts site surveys of potential TACFAC locations and documents the commander's critical information requirements (CCIRs) and other vital information. After the site survey is complete, the results of the CCIRs are shared, distributed, and coordinated with all pertinent units.

2-7. As the SFOD develops a facility construction plan, the commander may also conduct an HN capability assessment. In particular, the commander gets answers to the following questions:
- Are the HN soldiers able to assist the SF unit during the facility construction process?
- Are personnel identified to provide and deliver heavy equipment to the chosen location?
- Are construction materials available in country, and who receives compensation?
- Are large-capacity diesel-electric generators and spare parts available for local purchase?
- Are HN heavy equipment operators available for hire and how are they to be paid?

2-8. The site survey team deploys with sufficient portable electronic data-collection and data-storage devices (for example, computers, cameras, hard drives, and discs). When the SFOD mission is identified as particularly exceptional, individuals with area-specific qualifications (subject-matter experts [SMEs]) may deploy with the site survey team. Because building a new SF TACFAC in an austere environment is a unique endeavor, survey teams should include at least one Special Forces engineer sergeant (MOS 18C) as an SME.

2-9. Although most thorough survey missions are lengthy because they require answers to detailed questions, some survey missions may require only limited TACFAC development criteria and basic site selection. On these survey missions, SFODs preview potential sites, make appropriate choices, and return to base.

2-10. A comprehensive site survey requires the SFOD—particularly the 18C—to perform a complete review of all potential HN permanent structures and any available bare-ground locations that may require improvement. If a bare-ground site is selected, the initial site survey must include a tentative construction plan and a tentative bill of material. This plan includes any heavy construction vehicles, equipment, and personnel needed to build the TACFAC and addresses facility design, logistic accessibility, topographic layout, elevation, drainage, soil excavation, electric, plumbing, defense, and security considerations. The plan provides a tentative construction timeline that allows for the overall TACFAC construction stages to mesh together seamlessly.

2-11. A primary site survey consideration is the provision of seamless protection and security procedures for the SF TACFAC throughout each phase of construction. The SFOD determines who will provide this protection and how many personnel will be needed.

2-12. For a thorough survey, the SF unit incorporates an extensive and comprehensive resource review. This review may include verbal, written, and electronic media examples of lessons learned, unit after-action reviews (AARs), and SORs used by other SF units. If possible, Soldiers may speak with other SFODs and senior SF unit leaders regarding previous TACFAC missions. Most unit AARs and reports from the Center for Army Lessons Learned (CALL) are UNCLASSIFIED. (Appendix B contains a site survey format.)

AREA ASSESSMENT

2-13. The initial area assessment begins very early in the MDMP—immediately after mission receipt. The principal area assessment effort is ongoing and continues to be updated after the SF unit arrives in country. This assessment forms the foundation for a large part of the SFOD's subsequent activities in their AO. (Appendix C provides an area assessment format.)

SPECIAL OPERATIONS DEBRIEF AND RETRIEVAL SYSTEM

2-14. The SODARS is designed to collect debrief reports from SO personnel to be used as electronic reference material by the greater SO community. SF units use SODARS to retrieve detailed information about any SF TACFAC that may have been used by previous SF units. SODARS reports are classified documents housed on the United States Army Special Operations Command (USASOC) SECRET Internet Protocol Router Network (SIPRNET).

SECTION II—DESIGN

2-15. The preferred design option for an SF TACFAC in a hostile environment is hardened, independent, and capable of supporting and protecting the SFOD and HN soldiers in the worst-case scenario—the TACFAC coming under siege. The facility is built around an inner perimeter OPCEN or tactical operations center (TOC). Buildings and support infrastructure house the SFOD and possibly other government agencies (OGAs), and HN soldiers with or without their dependents.

> **Vietnam Tactical Facility Design**
>
> Square, triangular, pentagon and freeform TACFAC exterior designs were constructed by the early SF units in Vietnam. By 1968, there was a standard TACFAC interior design in use for buildings, firing positions, and mortar pits.
>
> In 1962, Camp Nam Dong was at first built in a freeform shape which, after being overrun, proved to be an ineffective design. In 1964, it was rebuilt into the classic triangular shape.
>
> Camp Gia Vuc was pentagon-shaped and one of the most menacing in Vietnam. It had more than 20 individual fighting positions, 2 machine gun bunkers per wall, 1 machine gun at each corner bunker, 18 mortar pits, 6 105-mm howitzers, and 2 155-mm howitzers, for a total of 15 direct-fire and 26 indirect-fire guns.

2-16. Each SF TACFAC is similar in function but unique in its design because it is dependent upon METT-TC, terrain analysis (OAKOC), and the material available in the AO. There are seven sectors or areas common to many permanent SF TACFACs. SF TACFACs in urban environments may be unable to accommodate all seven sectors. Also, many mature rural TACFACs incorporate shooting ranges; detention facilities; DFACs; gymnasiums; showers; laundry facilities; motor pools; morale, welfare, and recreation (MWR) facilities; HN administration and meeting rooms; HLZs; and airfields within the seven sectors. Beginning in the center of the TACFAC and moving outward, the seven sectors or areas that should be common to all SF TACFACs are best displayed in the example of permanent rural TACFACs depicted in Figures 2-1 and 2-2, page 2-4. The sectors or areas illustrated in Figures 2-1 and 2-2 are as follows:

- Inner perimeter (item A).
- Inner barrier (item B).
- Outer perimeter (item C).
- Outer barrier (item D).
- Administration area (item E).
- Access road (item F).
- Surrounding area (item G).

INNER PERIMETER

2-17. The inner perimeter is the heart of the SF TACFAC. Operational, administrative, and logistic operations are controlled from various hardened and protected buildings within this sector. At the very least, the SF unit should build one primary observation tower within the inner perimeter. Additional towers may be constructed as the TACFAC develops. The inner perimeter is vital to the defense of the TACFAC; it should be surrounded entirely and protected by an inner protective berm or wall.

Chapter 2

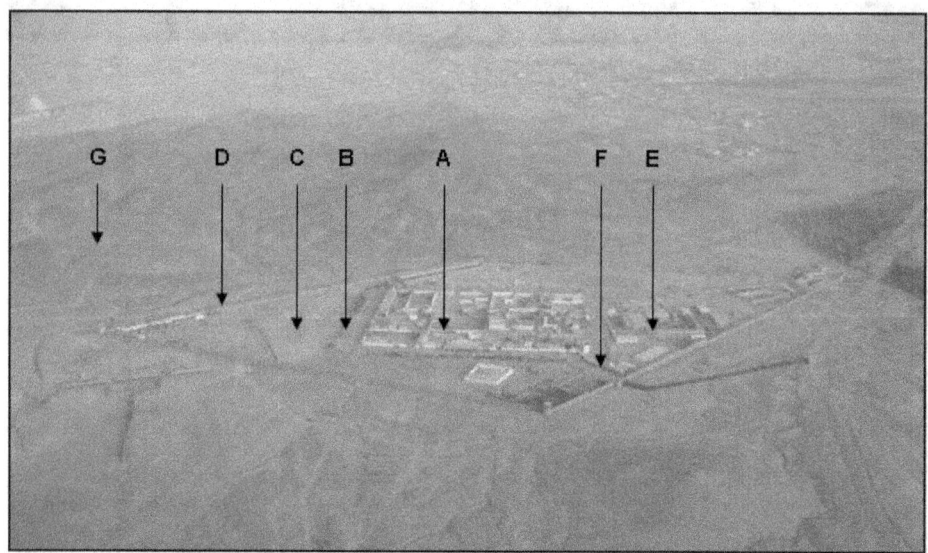

Figure 2-1. Rural Special Forces tactical facility overview (example 1)

Figure 2-2. Rural Special Forces tactical facility overview (example 2)

INNER PROTECTIVE BERM OR WALL

2-18. An inner protective berm or wall (Figure 2-3) is built from the ground up, and its construction should be one of the first priorities of the occupying SF unit. The protective inner berm may be constructed using available native materials, such as mud, sand, or lumber. The berm incorporates elevated bunkers and fighting positions for crew-served and individual weapons. All positions are shielded from the rear by a low-splinter wall, creating, in effect, an aboveground trench. By elevating the inner-berm fighting positions (Figure 2-4), Soldiers firing from the center of the TACFAC have clear fields of fire over the heads of friendly forces in the lower elevation of the outermost sectors. This elevation layering permits the TACFAC to maximize all defensive firepower.

Figure 2-3. Inner protective berm

Figure 2-4. Elevated inner berm positions

Chapter 2

PREDETONATION MATERIALS AND FRAGMENT SHIELDING LAYERS

2-19. The close proximity of a TACFAC to hostile urban or rural areas denies it the ability to control an adequate standoff area beyond the TACFAC perimeter. The concentration of operational and support billeting areas are a concern due to their high casualty potential and inherent lack of effective protective measures. Soldier's quarters generally are soft targets and usually consist of tents, trailers, and field-expedient structures. The soft nature of these quarters and the density of personnel render them vulnerable to fragmenting munitions such as rockets, artillery, and mortars (RAM).

2-20. The use of retrofit cyclone fencing and corrugated roofing designed as predetonation material and fragment-shielding layers (stand-off protection) in shelter areas and outer and inner perimeter defense can significantly enhance TACFAC survivability. Cyclone fencing can be used along the outer wall of a TACFAC in order to predetonate rockets prior to them entering the perimeter and detonating on a building or bunker or in close proximity to personnel. However, due to the number of structures involved and the square footage requirements, predetonation materials and fragment-shielding layers must be prioritized, cost-effective, and easy to construct at the TACFAC. Figure 2-5 is a three-photo sequence that illustrates the preemptive use of fencing material to predetonate a rocket.

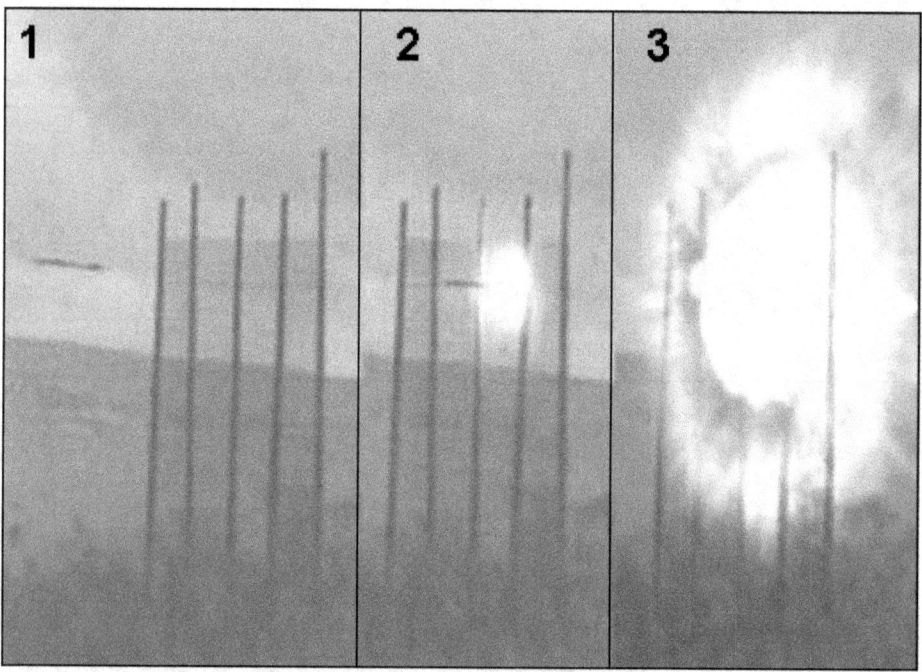

Figure 2-5. Rocket predetonation sequence

2-21. Chain-link fencing or durable wire mesh is installed in front of the firing ports of each building or bunker at an appropriate distance (it must be measured and tested) to disrupt, predetonate, and defeat rocket-propelled grenades (RPGs) and thrown explosive satchels or grenades. Fence sections are placed at an angle to prevent oblique shots from penetrating. If only a limited supply of fencing and sturdy wire mesh is available, the SFOD must prioritize by buildings and bunkers. Priority RPG fencing should be placed on the outer perimeter bunkers first, followed by inner perimeter bunkers, and then buildings (when available).

Tactical Facility Planning and Design

2-22. Various combinations of materials and techniques are being used that provide increased protection against blast and fragmentation effects. By adding effective blast-resistant walls, predetonation cyclone fencing, and fragment-shielding corrugated roofing over existing TACFAC structures, the potential for casualties is greatly diminished.

FIGHTING BUNKER MATERIALS AND DESIGN

2-23. If suitable local building materials are unavailable to build fighting bunkers and buildings, materials may be obtained through the USG or civilian contractors. Commonly used materials include stackable barrier system containers, containers express (CONEXs), concrete, Cinva-Ram blocks, 55-gallon drums, and sandbags. Large fighting bunkers may have sleeping areas; however, space is limited in even the most sizeable of bunkers. Sleeping bunks may be double or even triple stacked. Bunker floors are built to include drainage channels and grenade sumps. Whenever possible, all buildings and bunkers should utilize a dual roof or an improved protective system to mitigate the destructive effect of rocket and mortar rounds.

2-24. The main heavy weapon used in fighting bunkers is determined by METT-TC analysis. The main heavy weapons usually are medium machine guns, heavy machine guns, or automatic grenade launchers. Within each fighting position, the SF unit should keep or have immediate access to a supply of—

- Main-gun ammunition.
- Fragmentation grenades.
- Smoke grenades.
- White phosphorous (WP) grenades.
- Food and water.
- First-aid supplies.
- Communications equipment and laser rangefinder or compass (with extra batteries for electronic components).
- Night vision goggles (NVGs) and thermal imaging equipment (with extra batteries).
- Range card (Figure 2-6, page 2-8).

SPECIAL FORCES TEAM HOUSES

2-25. SF team houses can function as logistics centers, administration centers, isolation facilities (ISOFACs), and sleeping quarters for the SF unit. SF units must keep at least one (and preferably two) U.S. Soldiers awake during any 24-hour period. When eating, resting, or sleeping, SF personnel avoid bunching up. Wherever possible, Soldiers disperse and sleep in different buildings. Soldiers must know where to go during an enemy attack and the quickest route to their assigned positions. TACFAC visitors must also be briefed on appropriate emergency procedures. SF team houses, like fighting positions, are built above or below ground level. Aboveground structures should, at a minimum, be lined with sandbags stacked 2 rows wide and 10 rows high. A dual roof is constructed using two sets of corrugated metal sheets and sandbags. This creates a sandwich-like layer of metal sheeting and sandbags. The second roof provides weather protection, helping to keep sandbags dry and preventing them from soaking up excess water weight. Additionally, the extra material helps deflect and absorb kinetic-energy blasts. Bunkers close to the team house (Figure 2-7, page 2-9) provide additional protection during indirect-fire attacks.

OPERATIONS AND COMMUNICATIONS BUNKERS

2-26. The OPCEN and communications-electronics (CE) bunkers are the focal point structures of the inner perimeter. These bunkers typically are positioned belowground, have dual roofs (like the team houses) or extensive overhead protection, and two secure access points.

2-27. The CE bunker contains rooms for the SF unit and its HN counterparts. Entrance to the OPCEN bunker, however, should be limited to U.S. and vetted HN personnel only. Both OPCEN and CE bunkers are equipped with—

- Separate contingency generators to provide power for communications equipment.

Chapter 2

- Spare batteries for the lighting systems. The lighting is connected to the contingency generator only if it is large enough to sustain both the communications equipment and lighting systems. Batteries or solar power can be used as an emergency power source.
- Small-arms with additional ammunition.
- Fragmentation, smoke, and WP grenades.
- Food, water, and first-aid supplies.

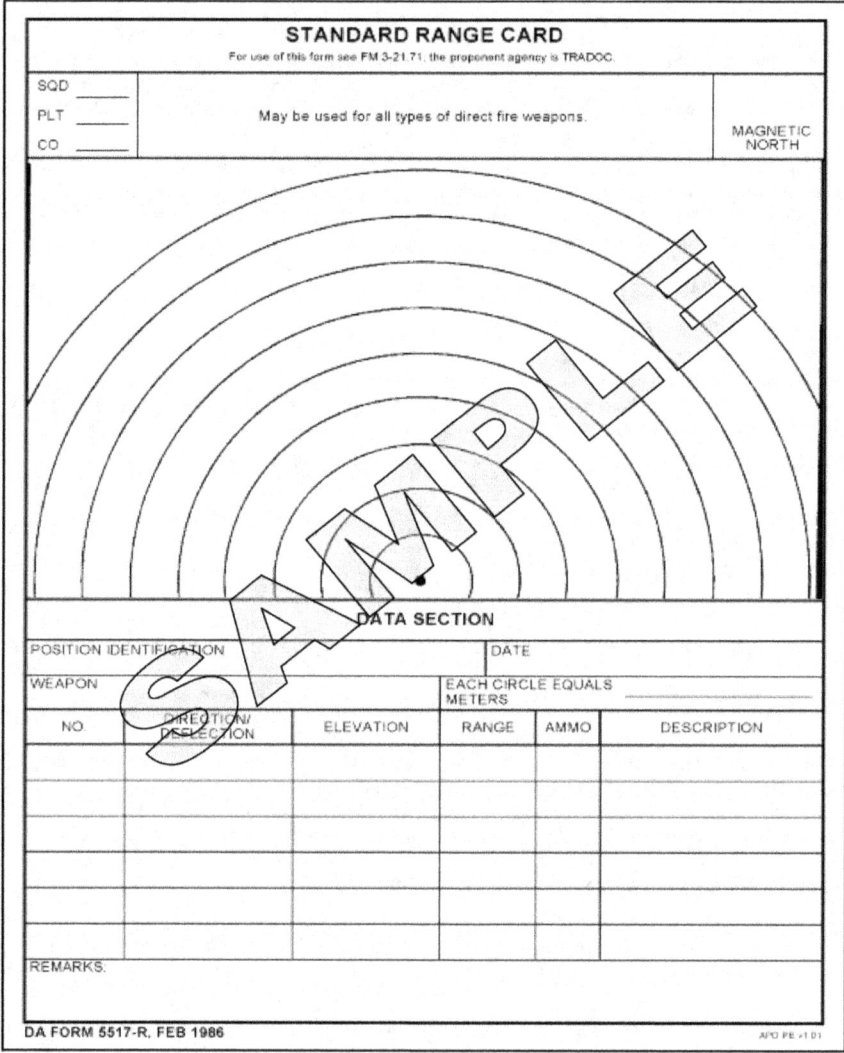

Figure 2-6. DA Form 5517-R (Standard Range Card)

Tactical Facility Planning and Design

Figure 2-7. Personnel bunker offering indirect-fire protection

SPECIAL FORCES MEDICAL BUNKER

2-28. The SF medical bunker should be underground and serves as the de facto field hospital only during TACFAC defense operations. Routine medical problems are treated in the main dispensary (discussed in paragraph 2-81), located in the administration area.

2-29. The 18D is responsible for the operation and maintenance of the medical bunker. The medical bunker should be large enough to house an operating room, ward room, recovery room, triage area, and medical storage area. The medical bunker stores a large supply of pharmaceuticals (to include narcotics). Access needs to be controlled by the 18Ds.

2-30. The medical bunker has a dual roof (like the team houses), and the bunker entrance points are large enough to allow access by aid and litter personnel. The bunker is stocked with adequate medical supplies, food, water, batteries, and ammunition, and is equipped with an emergency battery-powered lighting system. Litter racks may be constructed in the ward area to accommodate triple-stacking of patients awaiting or recovering from treatment.

GENERATOR BUNKERS

2-31. Although the diesel-electric generator bunkers may be built belowground, they normally are built aboveground with a dual or reinforced roof. This is because the cramped space of belowground generator bunkers makes even simple repairs and maintenance more difficult. When built aboveground, generator bunkers are dispersed and well fortified with sandbags. Generator bunkers incorporate a few feet of space between the top of the sandbag walls and the bottom of the roof to allow heat to escape and maintain proper air flow.

Chapter 2

2-32. There must be a minimum of two generators on site; one generator is operational while the other generator is in reserve, repair, or maintenance mode. Diesel fuel for electric generators is stored safely away from other flammable material. Chemical fire extinguishers are tested for operability and kept close to the generator bunkers. All high-voltage power lines are buried deep enough to address potential safety concerns. However, the lines must not be buried so deep that a damaged line cannot be located, unearthed, and quickly repaired.

AMMUNITION BUNKERS

2-33. Ammunition storage bunkers may be built above or below ground level. Belowground construction must be avoided in areas where there is standing water or where the groundwater table is high. When built aboveground, ammunition bunkers may be integrated into the inner-perimeter berm. For example, a concrete structure or CONEX-type container covered on three sides by the inner berm or sandbags provides good protection and access. A dual roof provides more protection for the ammunition bunkers from indirect fire.

2-34. The Special Forces weapons sergeant (MOS 18B) ensures that ammunition bunkers are adequately stocked and secured. All ammunition is stored aboveground on standard wooden pallets. A diagram or picture identifying where all the ammunition is located should be posted on the inside of the bunker. High-explosive (HE), WP, and illumination mortar ammunition are stored in quantity in the ammunition bunkers. All ammunition containers are sorted and palletized according to type and caliber to make the round type easy to identify. To open the ammunition containers, a tool box that includes a hammer, pry bar, and flashlight is kept in the bunker. The ammunition bunkers are kept locked; however, the Soldier responsible for the bunker must ensure that all SF Soldiers have the keys or tools needed to gain immediate access to the ammunition.

MORTAR POSITIONS

2-35. Mortar positions located inside the inner perimeter may be above or below ground level. For optimum protection, mortar pits are built belowground, and the interior is revetted using concrete or a minimum of two layers of sandbags. The position may be strengthened further using standard 55-gallon drums or steel howitzer-propellant shipping containers filled with earth. A tarp should be used to protect the mortar from the elements. The mortar position (Figure 2-8, page 2-11) must allow for a clear, 360-degree field of fire.

2-36. The 18B, in coordination with the designated mortar gunners, ensures that an assortment of mortar rounds is prepared for rapid response. Ready-racked ammunition includes HE quick rounds with the propellant charges cut for preplanned defensive fire missions. Illumination rounds have the time-delay fuse preset and the charges cut. Ammunition stocks are rotated; the oldest rounds are used first.

2-37. Because metal ammunition cans cause unintended sparks, wooden ammunition crates are used to store the accumulation of unused flammable mortar powder bag propellant charges. A chemical fire extinguisher must be readily available.

2-38. A call-for-fire checklist is established for mortars and communication equipment. The senior 18B coordinates with the senior Special Forces communications sergeant (MOS 18E) to ensure all communication equipment functions correctly. Communications and firing procedures must be rehearsed.

SUPPLY AND ARMS ROOM

2-39. The dual-use supply and arms room is the responsibility of the 18B and the 18C. This room typically is built aboveground and protected with a double layer of chest-high sandbags or material of equal protection. A dual roof is used to protect the facility from indirect fire. The structure is big enough to house supplies and equipment in one section and small arms and crew-served weapons in another. Weapons storage racks are used to sort weapons by type, caliber, and frequency of use.

Figure 2-8. Mortar position

2-40. The supply and arms room (particularly the arms room section) must be properly secured at all times. One technique used to secure the area is to provide HN and SF sleeping quarters within the building and rotate guard responsibilities using a duty roster.

VEHICLE REVETMENTS

2-41. Vehicle revetments are used as hull-down covered and concealed parking bays for tactical combat vehicles. One common construction technique requires earth removal and build-up protection using the displaced earth and 55-gallon drums. However, revetments also are constructed using stackable barrier systems. Revetments are dispersed throughout the inner perimeter. Tactical vehicles may be backed into the positions so that they can be moved rapidly in emergencies. Depending upon the tactical situation and the amount of space available, SF units may include primary and alternate revetments for armored vehicles and artillery pieces.

LATRINES

2-42. The senior 18D is accountable for the construction of a sufficient number of latrine sites throughout the TACFAC sectors. Latrines should be placed near shower locations and an appropriate distance from living and working areas. The number and size of the latrine facilities vary according to the number of personnel operating in the TACFAC. The 18D may construct a field-expedient burn barrel latrine (solid and liquid waste-disposal system) using 55-gallon drums.

Chapter 2

TRASH POINT

2-43. Trash generated in the inner perimeter can be burned in 55-gallon drums. Trash from the other sectors should be burned in the outer perimeter (or beyond) to prevent the spread of disease. Fire extinguishers must be available wherever trash is burnt.

EMERGENCY SIGNALS

2-44. An emergency signal technique, such as a fire arrow, is constructed in the inner perimeter to signal support aircraft. When all other forms of electronic communications and navigation systems fail, the fire arrow allows aircraft to rapidly acquire the direction and distance to the emergency.

FIRE POINTS

2-45. Fire extinguishers, firefighting stations, and smoke detectors are positioned throughout the TACFAC. The senior 18C has the responsibility to emplace and maintain chemical fire extinguishers at firefighting stations throughout the facility. Fire extinguishers must be available for all vehicles, occupied buildings, and hazardous equipment, fuel, and ammunition storage areas.

2-46. When available, a good fire-control asset is a water tanker truck or trailer with adequate pumping capacity, sufficient water and hoses, and a trained firefighting team centrally located within the inner perimeter. A man-portable wheeled fire-containment system may suffice in smaller TACFACs. Firefighting stations are fabricated using 55-gallon drums and buckets.

INNER BARRIER AND OUTER BARRIER

2-47. The purpose of the inner and outer barrier is to slow the adversary during a breach and prevent them from exploiting their offensive action. As their attack slows, the enemy incurs more casualties from the protected fighting positions of the TACFAC. The longer it takes the enemy to negotiate either barrier, the more casualties they will sustain. As such, the barriers permit a small defensive force to successfully defend against a numerically superior attacking force.

2-48. There should be two passages leading through the barriers. The first passage is an access road that is used by vehicles entering the inner perimeter. The second passage is a personnel path that leads from the inner perimeter to the outer perimeter gate. Both passages must be controlled at all times by manned guard posts.

2-49. The barriers normally have two or more rows of barbed-wire entanglements. The intervals between the rows also contain additional wire obstacle belts. All wire fences or barriers may be rigged with trip flares and booby traps to provide early warning to the TACFAC. The barriers channel enemy forces to create casualties and impede movement. The barriers are covered with direct-fire weapons from positions on the inner and outer perimeter.

2-50. The barriers may be constructed using a variety of obstacles, including double-apron barbed-wire fencing, mine belts, tanglefoot, triple-strand concertina wire, canalizing fences, trip flares, and trenches. The double-apron barbed-wire fence is approximately 1 meter high and 3 meters wide, and it is reinforced with additional wire. Field-expedient noisemakers (such as rocks or pebbles in empty tin cans) may be attached along the length of the barbed-wire fence at various heights.

Note: Additional barrier construction techniques are included in FM 5-34, *Engineer Field Data*, and Soldier Training Publication (STP) 31-18C34-SM-TG, *Soldier's Manual and Trainer's Guide, MOS 18C, Special Forces Engineer Sergeant, Skill Levels 3 and 4*.

PERIMETER ROAD

2-51. The perimeter road is a utility road that encircles the TACFAC along the inside of the outer barrier. It is used primarily to transport personnel, material, and equipment to service construction and maintenance projects. It often is safer and easier to approach maintenance and repair problems from inside the TACFAC. However, during TACFAC defense operations, the perimeter road within the security envelope of the TACFAC provides a rapid and protected route to quickly move men or materiel within the facility.

ANTIPERSONNEL AND ANTITANK MINES

2-52. The SF unit must exercise caution if antipersonnel (AP) or antitank (AT) mines are incorporated into the barrier defense plan. The location, marking, laying, recording, and removal of mines are covered in detail in STP 31-18C34-SM-TG. The SF unit must record information on the location of each mine in the minefield using Department of the Army (DA) Form 1355-1 (Hasty Protective Row Minefield Record). Recordkeeping requires (at a minimum) an accurate sketch and compass reading; more accurate measuring techniques include digital photos, laser rangefinder readings, and global positioning system (GPS) data.

2-53. Because of the danger posed by AP and AT mines to friendly forces and HN civilians, the legality of employing mines as part of the TACFAC defense, and the restrictions imposed by rules of engagement (ROE), their use must be weighed carefully. Under the 1996 Protocol on Prohibitions or Restrictions on the Use of Mines, the U.S. Armed Forces may use claymore mines in command-detonated modes. A 1997 Mine Ban Treaty placed additional restrictions on the use of mines; however, the United States is not a signatory of this treaty. ROE guidance provided by higher SF headquarters (HQ) must be followed regarding mine employment.

2-54. Command-detonated AP claymore mines are very effective in TACFAC defense when allowed by the ROE. Claymore mines should be staggered and placed in a minimum of two rows and fired sequentially against assaulting forces. The outermost row is fired first, followed by the second outermost row (and so on). In order to make the most efficient and effective use of claymore mines, the SF unit—

- Selects mine locations carefully, marking and emplacing the mines only one time.
- Emplaces mine rows at least 3 meters apart to prevent sympathetic detonation.
- Verifies that each mine interlocks fire with the neighboring mines.
- Ties knots at the end of the firing wire to help indicate which row it fires. For example, the first-row firing wire should have one knot, the second-row firing wire should have two knots (and so on).
- Instructs Soldiers to not waste claymore mines on targets that may be better defeated by other means, such as mortars or crew-served weapons.
- Establishes a prearranged warning signal to alert outer-perimeter personnel that inner-barrier claymores are about to be fired. This will enable outer-perimeter personnel to seek cover until after the mines are fired.
- Disconnects the M-57 firing devices (commonly referred to as "clackers") each day, replaces all electrical covers to prevent corrosion and debris, and reconnects all devices each evening.
- Limits each firing bank to five or fewer claymore mines.

FLARES

2-55. Trip flares are attached to all wire fences of the barriers. Flares provide two key benefits to the occupants of the TACFAC. First, they serve as early-warning devices, alerting the SF unit to potential security breaches. Second, they provide illumination to help identify and engage potential targets. Flares are prone to deterioration, and they must be inspected, maintained, and replaced regularly.

2-56. Trip flares also are attached to a separate solid metal or wooden stake located on the friendly side of the wire. Trip wires run forward—through the obstacle—to the outer enemy side of the wire. This technique makes it more difficult for the enemy to disarm the flares.

TANGLEFOOT

2-57. Tanglefoot is employed to disrupt enemy assaults and sappers. Tanglefoot fields may be 10 meters deep and staked at irregular intervals between 1 to 2 meters. Barbed wire is strung from these stakes in an irregular crisscrossed pattern 6 to 8 inches above the ground. Positioning wire at this height prevents an

Chapter 2

assaulting force from gaining momentum. Furthermore, crawling adversaries need to rise up to climb over the wire, thereby presenting an easily engaged target.

TRIPLE-STRAND CONCERTINA WIRE

2-58. Triple-strand concertina wire is positioned in a coiled fence that is approximately 2 meters high and 2 meters wide. The SF unit installs the concertina rows using taut horizontal wire to keep rows secure and prevent the adversary from depressing the coils. The bottoms of the coils are staked to the ground every 5 to 10 feet.

CANALIZING FENCES

2-59. Canalizing fences are constructed using triple-strand concertina wire. These fences crisscross all wire fences, entanglements, and obstacle belts of the outer barrier. The purpose of the canalizing fences is to slow and funnel the enemy into the path of heavy automatic weapons fire.

Note: Detailed instruction for constructing fences using triple-strand concertina wire may be found in FM 5-34.

TRENCHES

2-60. Trenches provide cover and concealment for friendly personnel moving throughout the TACFAC. They may be used by personnel as a line of communication (LOC) to carry litters, messages, or ammunition. Trenches provide a protected route for personnel movement during battle. The time, effort, expense, and maintenance required by a particular trench must be weighed to determine its overall value.

2-61. Trenches cross through the inner barrier to the outer perimeters in a random, zigzag pattern. However, the outer perimeter trench completely encircles the inside of the outer perimeter and connects all the defensive positions. All trenches need to be at least 6 feet deep and wide enough to permit two-way foot traffic and litter carry. Retaining walls are used to prevent erosion and cave-ins caused by adverse weather. Additional information regarding trench construction may be found in FM 5-34.

OUTER PERIMETER

2-62. The outer perimeter contains key TACFAC defensive positions and most of the HN force. Layers of defensive fighting positions, bunkers (Figure 2-9, page 2-15), and buildings are located in the outer perimeter. Outer-perimeter features include fighting bunkers, HN force housing, sleeping bunkers, ammunition bunkers, observation towers, vehicle revetments, mortar positions, company command-post bunkers, latrines, water points, firefighting points, a DFAC, a motor pool, a fuel point, and a perimeter trench system. The outer-perimeter trench connects the sector fighting positions, command posts, mortar pits, and ammunition bunkers with the trench system of the inner perimeter. The outer perimeter typically has at least a 25-meter range and small-arms test-fire area.

FIGHTING BUNKERS

2-63. Outer-perimeter fighting bunkers should be constructed belowground, thereby allowing inner-perimeter bunkers (at a higher elevation) to engage targets by firing over the heads of friendly forces. The bunkers should be at least 6 feet deep. Logs or other building material used for the sides and roof should be at least 6 inches thick. Each fighting bunker should contain at least one automatic weapon, ammunition, grenades (fragmentation, smoke, and WP), first-aid kit, food, and water. All positions should have range cards, sector sketches, and photographs of their assigned target areas. A section of wire fencing should cover the firing ports to defeat RPG rounds and thrown grenades. Additional information on bunker construction may be found in FM 5-34.

Tactical Facility Planning and Design

Figure 2-9. Machine gun bunker

HOST-NATION FORCE HOUSING

2-64. The HN-force outer-perimeter housing may be built either above or below ground level. Many HN construction projects, to include housing, mirror those of the SFOD when funding and construction time are available. Housing built belowground using CONEX-type containers expedites the construction process and requires fewer sandbags. When built aboveground, a double layer of sandbags—stacked at least chest-high—is required (Figure 2-10, page 2-16).

HOST-NATION FORCE SLEEPING BUNKERS

2-65. Each of the HN-force fighting bunkers has an adjacent sleeping bunker. The sleeping bunker serves as the living quarters for the HN force staffing that particular section of the outer perimeter. The fighting and sleeping bunkers make up a single bunker that is connected by a perimeter trench passing through the center. An L-shaped entrance with steps should be positioned at the rear of each sleeping section to allow traffic between the bunker and perimeter trench.

AMMUNITION BUNKERS

2-66. Belowground ammunition bunkers should be constructed on the outer perimeter, with the doors facing to the inside of the TACFAC. Most four-sided TACFACs have four ammunition bunkers because when ammunition bunkers are dispersed, they facilitate distribution. More importantly, bunker disbursement prevents any single bunker explosion from depleting all TACFAC ammunition. The bunkers should be positioned close to the intersection of the communications trench and the outer perimeter, allowing easy access by personnel of the HN-force defensive positions. Ammunition bunkers should have a double-wide door for easy movement of ammunition pallets and crates. A dual roof provides extra protection against indirect fire.

Chapter 2

Figure 2-10. Sandbags stacked chest-high around tactical facility structures

OUTER-PERIMETER TRENCH

2-67. The outer-perimeter trench inside the TACFAC encircles and connects all the defensive positions of the outer perimeter. The trench system should be deep and wide enough for two-way foot traffic and litter carry by two SF Soldiers. Adequate drainage must be considered during the initial planning process. Each section of the trench contains numerous individual fighting positions with firing ports positioned to cover the outer barrier, firing step-ups, and grenade sumps. The rear wall of the perimeter trench can also serve as an alternate command post in the event that the primary command post is destroyed.

OBSERVATION TOWERS

2-68. Observation towers enhance TACFAC defense when emplaced around the outer perimeter. These towers are designed to increase the visibility for the SF unit and to facilitate coordination for interlocking fires. A square rural permanent TACFAC should employ at least five guard towers—one in the center and one at each corner (Figure 2-11, page 2-17). Depending on the threat and the environment, employment of sniper blinds may be used to restrict the enemy's view of the inside of the perimeter guard towers. The bases of the towers should be enclosed with concertina wire. To maintain secure access, a ladder should be manually raised and lowered by the tower personnel. The guard tower includes a ready supply of—
- Main-gun ammunition.
- Fragmentation, smoke, and WP grenades.
- Food and water.
- First-aid supplies.

- Communications equipment and laser rangefinder or compass (with extra batteries for electronic components).
- NVGs and thermal imaging equipment (with extra batteries).
- Range card.

Figure 2-11. Outer perimeter corner observation tower

VEHICLE REVETMENTS

2-69. Outer-perimeter vehicle revetments are constructed in the same manner as the inner-perimeter vehicle revetments. The revetments are built parallel to the perimeter trench to afford vehicle protection from both friendly and enemy fire. Vehicles may back into defensive positions to allow turret weapons to be used during attacks.

MORTAR POSITIONS

2-70. Although light mortars may be employed in the outer perimeter, most TACFACs employ 81-millimeter medium mortars and 120-millimeter heavy mortars. As with the inner-perimeter mortar positions, outer-perimeter mortar positions are built belowground and have a supply of ammunition nearby. Adjacent bunkers with overhead cover serve as crew shelters and ammunition storage bunkers.

ALTERNATE COMMAND-POST BUNKERS

2-71. Belowground alternate command post bunkers can be built into the rear wall of the outer perimeter trench. In a square-shaped TACFAC, there may be up to four alternate command post bunkers—one per side. Radios and telephones are used to link the alternate command posts to the inner perimeter TOC.

Water Point

2-72. A TACFAC water point, such as a safe and secure well, is central to the survival of SF and HN personnel. During original construction of the facility, the plans for a potable water system must center on the worst-case scenario—the TACFAC coming under siege. Emergency water storage tanks are built at and above ground level throughout the TACFAC. Using pumps, filters, and purification systems, the water point provides potable running water in as many buildings as possible. A few designated aboveground structures serve as shower points for SF and HN personnel.

Fire Points

2-73. Firefighting equipment, such as extinguishers and buckets, are positioned throughout the outer perimeter. The construction of firefighting stations and placement of extinguishers is similar to those of the inner perimeter.

Dining Facility

2-74. The aboveground DFAC includes a dining area, kitchen area, food-processing area, food-storage area, and food-service line. The 18D and HN medical personnel monitor the food supply and food-preparation procedures. Whenever possible, U.S. personnel make every effort to eat and drink the same ethnic food and beverages as the HN personnel.

Motor Pool

2-75. The outer-perimeter motor pool may be a building or a covered storage shed dedicated to the service and repair of vehicles. It has several rooms at one end used for parts storage and office space. Unserviceable vehicles are kept in the motor pool overnight. Serviceable tactical vehicles are positioned inside the outer-perimeter vehicle revetments at night. A unit SOP is developed to include weekly vehicle and equipment maintenance.

Fuel Point

2-76. The fuel point (Figure 2-12, page 2-19) is within the outer perimeter and is the main TACFAC storage area for diesel and gasoline. Fuel may be stored in approved fuel bladders, 55-gallon drums, or a fuel trailer away from other hazardous material. An appropriately sized enclosure completely surrounds the fuel point to contain hazardous liquids. Firefighting equipment and fire extinguishers are located nearby. Smaller reserve fuel-storage enclosure areas may be positioned near the primary and alternate diesel-powered electric generators. All tactical vehicles are topped off with fuel each time they return to the TACFAC. All other vehicles and equipment are fueled according to unit SOP.

Trash Point

2-77. An outer perimeter trash point is constructed a safe distance from the fuel point. Trash is burned away from occupied building and bunkers; however, it typically is located within the TACFAC to control personnel access and hazardous material.

Weapons Range

2-78. The outer perimeter usually is large enough to support a covered small-arms test-fire area and 25-meter range. If space is available, a long-distance range (1,000 meters or more), with shooting positions along the inner barrier berm, provides sustainment training for all HN and SF Soldiers. Prior to every mission, rifles, pistols, and machine guns must be test fired.

Tactical Facility Planning and Design

Figure 2-12. Fuel point

ADMINISTRATION AREA

2-79. A separate room or building made available for HN meetings facilitates TACFAC security. The TACFAC administration area allows SF personnel to maintain close contact with the local populace within their AO. Communication with the local populace provides the SFOD with human intelligence (HUMINT), thereby increasing their security during combat and reconnaissance patrols. The reciprocity between the SFOD and the local populace ("winning hearts and minds") results in reduced enemy activity and greater security for the civilian population. This interaction allows the SFOD eventually to hand over the fight to the HN forces, defeat the insurgency, and return home.

2-80. The administration area is normally located between the outer barrier and the airfield, close to the surrounding area. The proximity of the administration sector to the TACFAC outer barrier depends upon METT-TC and HN support. Regardless of the distance, the SF unit maintains close contact with local HN personnel by offering medical, veterinary, and administration services in a secure location. Other key features of the administration area include the HN dispensary, the HLZ (Figure 2-13, page 2-20), an HN government building, the TACFAC main gate, and the main-gate bunkers.

DISPENSARY

2-81. An aboveground HN dispensary (Figure 2-14, page 2-20) may be located in the administration sector. Rows of sandbags—stacked two-deep and chest-high—around the outside of the dispensary provide protection for occupants against small-arms fire. The building may contain a waiting room, an office, a treatment room, and a ward room for patients that must stay overnight.

Chapter 2

Figure 2-13. Helicopters landing at a tactical facility helicopter landing zone

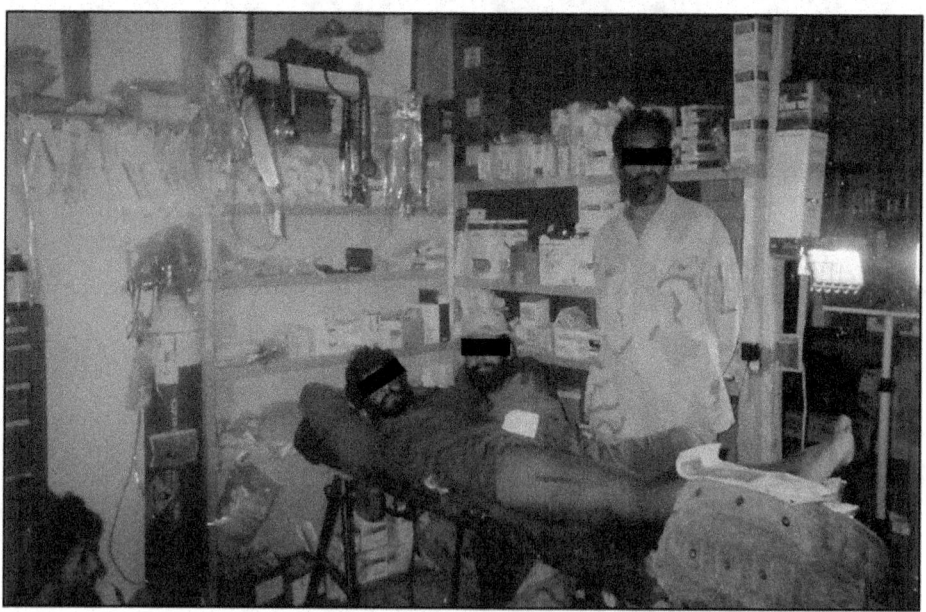

Figure 2-14. Host-nation dispensary

2-82. The HN dispensary provides around-the-clock medical care to the HN force and local HN civilian personnel. HN medics and nurses normally handle dispensary night-shift operations; 18Ds remain on call for emergencies that exceed HN medical staff capabilities. The senior 18D is responsible for determining if patients need to be evacuated to rear-area medical support facilities.

HELICOPTER LANDING ZONE

2-83. The HLZ provides a secure location inside the TACFAC for rotary-wing aircraft takeoff and landing. The concrete pad should be built large enough to accommodate at least one CH-47 Chinook or the new CV-22 Osprey. The surface of the pad must be solid enough to support aircraft in all types of weather; reinforced concrete is the preferred material (Figure 2-15). The pad surface is marked with a large letter H to indicate the HLZ. A windsock positioned nearby aids pilots during takeoff and landing. Aircraft unable to land at the secure TACFAC HLZ may be diverted to a larger (and potentially less secure) airfield in the surrounding area.

Figure 2-15. Constructing a rebar reinforced concrete helicopter pad

GOVERNMENT BUILDING

2-84. The primary function of the HN government building is to make HN government services available to the local populace and HN soldiers. Because medical treatment is a primary government service and has a direct impact on the local perception of the government, the government building should be positioned close to the dispensary. The government building, constructed aboveground, is protected from small-arms fire by a double-stacked row of chest-high sandbags around the outside of the building. The building is open to the public only during hours of daylight. The SF unit must consider the possible requirement of protecting the HN government building staff and locals. The unit must have a plan for them to take refuge in the TACFAC prior to enemy offensive operations.

Chapter 2

2-85. The SF unit may decide to house these employees in the TACFAC administration or other surrounding area secure buildings during hours of darkness. Distinctive items of clothing—such as colored vests (Figure 2-16)—may be used to identify trustworthy and vetted HN staff and employees. When there is no viable HN military force in the AO, local HN protection becomes the responsibility of the SF unit. If the AO is unstable and not secure, governmental services cannot continue. HN civilians are relocated to a separate and secure compound in a nearby larger village.

Figure 2-16. Vetted host-nation employees wearing colored vests

Main Gate

2-86. The main gate (Figure 2-17, page 2-23) controls entry to the facility via the single access road that runs through the TACFAC to the main gate. There may also be a second, smaller personnel-only gate alongside the main gate or at the opposite end of the TACFAC. All HN workers enter through this gate where they are searched and pick up their identification badges or passes. HN workers are required to return all badges and passes prior to leaving the facility. HN workers are inspected and searched. Workers must not be permitted to bring mobile telephones, cameras, or recording devices into the TACFAC. Employment of emerging biometric technology by the SFOD will ensure that local workers and visitors are properly screened and identified. All gates must be guarded around the clock. There should be a minimum of two guards at each gate to screen all personnel and vehicles. At night, an additional sergeant of the guard is on duty.

Tactical Facility Planning and Design

Figure 2-17. Main gate entrance and confinement zone

2-87. The material used to construct the gate is in direct correlation to the threat. The gate may be constructed of wood and barbed wire or steel and concrete. The gate opening must be wide enough to allow a single lane of vehicular traffic. Bollards may be used to enforce traffic safety and security.

2-88. The SF unit may construct a confinement zone for all vehicle and foot traffic. Using this technique, a single narrow lane is created with steel bollards and concrete barriers. The road bed is made of rocks and loose gravel to encourage slow-speed travel. Vehicles entering the TACFAC wind around a series of obstacles and bollards under constant observation by TACFAC security forces. Hidden command-detonated explosives or claymores may be positioned along the confinement zone.

MAIN-GATE BUNKERS

2-89. As least one (preferably two) bunkers may be constructed at each gate. The bunkers are equipped with automatic weapons and house electric and alternate manual controls to activate (raise) and deactivate (lower) the gate or security bollards. When activated, some types of bollards are capable of bringing a 25-ton truck moving at 50 miles per hour to an immediate stop.

2-90. Main-gate bunkers are constructed in a manner similar to those of the inner and outer perimeter. All gate bunkers are manned around the clock. Normally, two bunkers (at either end of the confinement zone) are used as traffic control points. The main gate entrance bunker controls the flow of local HN personnel that have official business in the TACFAC. One or more of the outer barrier bunkers may be used to prevent or limit access of local HN workers to unauthorized TACFAC areas.

ACCESS ROAD

2-91. The access road (Figure 2-18) is a secure single-lane road that passes through the TACFAC and connects to the main road outside the main gate. Because the access road breaches all critical TACFAC perimeters, obstacles, and barriers, special consideration must be given to its construction and security. Gates and guard posts are constructed along the road at each critical crossing point. Access road control may be enhanced by incorporating a number of guard posts, bollards, and confinement zones.

Figure 2-18. Access road

SURROUNDING AREA

2-92. The surrounding area is the largest of all the sectors and the most susceptible to attack. The area surrounding the TACFAC should contain cleared fields of fire, a perimeter road, a main road, and, if possible, an airfield. All civilian HN vehicles initially park in designated marked parking lots in the surrounding area. A nearby building is used as an HN meeting house to sort through and select vetted drivers, laborers, and other personnel on official TACFAC business. As personnel are cleared, they may continue to pass through the surrounding area and into the administration area.

CLEARED FIELDS OF FIRE

2-93. The SF unit creates cleared fields of fire by trimming or removing all grass, weeds, stumps, trees, and debris from around the entire TACFAC. When local lumber is harvested in the course of the clearing operation, it may be used or stockpiled for later use inside the TACFAC. After the area is cleared, SF Soldiers physically walk the ground and identify any dead space. The dead space is recorded on a range card, sector sketch, or photograph and is covered by the TACFAC indirect fire support, mortars, rockets, or artillery. SFOD members should walk the ground around the TACFAC to identify actual or potential TACFAC vulnerabilities, listing and then ranking them using the criticality, accessibility, recuperability, vulnerability, effect, and recognizability (CARVER) matrix. Each SFOD member should ask, "Where would I attack?"

2-94. The cleared fields of fire are as large as the SF unit can practically make them, extending at least beyond enemy rifle range (300 to 500 meters). Enemy forces may attempt to construct ambush sites, improvised explosive devices (IEDs), tunnels, or assault positions close to the TACFAC during hours of darkness. Prior to first light, the enemy camouflages their work and disperses. Moonless nights, periods of heavy rainfall, and thunderstorms are ideal weather conditions for adversary construction activities. These projects may continue for days, weeks, or even months. To mitigate the threat of enemy construction, TACFAC reconnaissance patrols must be conducted at irregular and random intervals and locations throughout the AO.

2-95. Cleared fields of fire also provide an ideal known distance (KD) shooting range. Small berms are built at various known distances and reactive steel targets provide instant feedback to the shooter.

Main Road

2-96. The main road or highway serves as the primary link between the TACFAC and other surrounding area towns and villages within the TACFAC AO. Normally, this road is heavily travelled by military and civilian vehicles.

Airfield

2-97. U.S. Army utility and cargo helicopters (such as the UH-60 Blackhawk and CH-47 Chinook) usually are required for high-priority operational missions and may be unavailable to deliver sufficient resupply to SF TACFACs. Many rural TACFACs—particularly the larger AOBs—will require an airfield so that United States Air Force (USAF) cargo aircraft can provide resupply. Enemy forces will quickly grasp the importance of the airfield to TACFAC survival. The SF unit must furnish adequate security for the airfield to prevent enemy disruption during vital resupply efforts.

This page intentionally left blank.

Chapter 3

Tactical Facility Construction

A rural TACFAC built today in a remote valley of Afghanistan is very different in construction materials than a TACFAC built in the dense jungles of Colombia. However, basic construction principles or standards remain the same regardless of TACFAC location. In Vietnam, SF experimented with a plethora of TACFAC materials and designs, but no particular TACFAC proved impenetrable.

CONSTRUCTION PRINCIPLES

3-1. All TACFACs have a similar purpose, and SFODs perform operations in a parallel manner regardless of the diverse environments. However, the TACFAC construction principles remain uniform for aboveground, ground-level, or belowground structures. Construction principles incorporate METT-TC, terrain analysis, available materials, water table, soil content, weather, and the construction abilities of the SFOD.

3-2. Most administrative buildings are built aboveground using wood frames with plywood floors and walls. In mild climates, solid walls may be built to hip-level, with wire-mesh screening forming the upper portions of the wall. Adequate sandbag protection must be considered. Perimeter operational structures used to house SF Soldiers or to store hazardous material, ammunition, and fuel should be protected and constructed belowground.

GENERAL STRUCTURAL COMPONENTS

3-3. The general structural materials employed to protect buildings and fighting positions depend on the weapons or effects they are designed to defeat. All buildings and fighting positions have the typical configuration of floor, walls, and roof designed to shield and protect equipment, materials, and occupants. The floor, wall, and roof components either support the shielding or make up that shielding. These structural components also must resist blast and shockwave effects from detonation of HE rounds.

FLOORS

3-4. Buildings and bunker fighting position floors may be made from almost any material, provided it resists weathering, wear, and traffic. Concrete slabs provide the most effective flooring surface. Soil or sand may be used if other resources are unavailable; however, they are least resistant to water damage, rotting, and foot and vehicle traffic. Wood pallets or other field-expedient materials may be cut to fit floor areas. Drainage sumps or drains are installed where excess water is a problem.

WALLS

3-5. Walls of buildings and fighting positions are of two basic types—belowground (earth or revetted earth) and aboveground. Belowground walls are made using the in-place soil remaining after excavation of the position. This soil will need revetment for support. The amount of support depends upon soil properties, drainage, and the angle and depth of cut. Aboveground walls are normally constructed to provide shielding from direct fire and indirect fragmentation. These walls are usually built of sandbags, concrete, or other local materials.

Roofs

3-6. Roofs of buildings and fighting positions are designed to support overhead cover as shielding from mortar fragments and small-caliber direct fire. Contact-burst protection requires much stronger roof structures and requires careful material selection and design. Roofs able to support adequate shielding are constructed of sturdy concrete or steel.

Machine-Gun Bunkers

3-7. Most machine-gun bunkers feature a single firing port for each machine gun and may be configured for up to three machine guns. Sandbags surrounding machine-gun bunkers are more durable when capped with concrete. The bunker may connect to a perimeter trench with an entrance that has at least one 90-degree turn (to protect against hand grenades). If multiple machine guns are employed in a single bunker, the fields of fire of the machine guns overlap.

3-8. A bunker with three machine guns provides an enormous amount of defensive firepower; however, there are a number of drawbacks to such positions. When three guns are employed in a single position, the size of the machine-gun bunker must be increased. This requires additional material and labor. Furthermore, three machine guns firing from a single position may potentially use excessive amounts of ammunition. This requires a larger ammunition storage area and demands more Soldiers. The machine guns make up a large percentage of the unit's overall firepower; if grouped together in such close proximity, the elimination of this single bunker may effectively cripple the TACFAC.

3-9. Excavation for a single machine-gun bunker produces a cube-shaped hole measuring 5 to 7 feet on all sides. The excavated dirt is used to fill sandbags to emplace around the bunker. The hole is revetted with planks, sandbags, or corrugated steel. Layered sandbags are used to create the walls for roof support. A 463L cargo pallet is used as the basis for the roof. On top of the pallet, an application of tar paper is used for additional waterproofing. Sandbags are stacked on top of the tar paper. For extra roof support and stability, vertical 4-inch by 4-inch posts may be placed along the inside walls of the bunker. The exterior sandbags of the emplacement can be capped in concrete to protect sandbags and lumber from deterioration caused by sunlight, rain, insects, and foot traffic.

Mortar Positions

3-10. The SF unit may have 60-millimeter, 81-millimeter, and 120-millimeter mortars at their disposal. Ground-level or, ideally, belowground mortar positions are constructed for each mortar. The SF unit constructs a pit by excavating a circular area, approximately 10 feet in diameter. The floor of the pit will remain bare earth to facilitate water runoff and allow Soldiers to properly set the mortar base plate. Because mortar aiming is based on the concept that 6,400 mils comprise a circle, the SF unit should paint or apply legible mil-number markings around the inside circumference of the pit with luminescent paint or tape (Figures 3-1 and 3-2, page 3-3). These markings facilitate placing faster and more accurate rounds on target.

3-11. An effective field-expedient and ground-level pit can be built using sand-filled 55-gallon drums or stackable barrier system-type barriers. The unit places the drums or barriers in a circle and stacks a double layer of sandbags around the inside and outside of the drums, creating an impenetrable small arms barrier. All exposed sandbags should be capped with concrete.

3-12. Regardless of the design used, the mortar crew needs an easily accessible entryway. An enclosed and protected ammunition supply bunker is attached to the mortar position. A supply of mortar rounds is kept readily available for emergency and immediate-use fire missions. These rounds include HE, WP, and illumination rounds.

Tactical Facility Construction

Figure 3-1. Mortar pit with mil markings

Figure 3-2. Mortar pit without mil markings

Chapter 3

OBSERVATION TOWERS

3-13. Observation towers (Figure 3-3) enhance TACFAC security and communications. The overall protection and security posture is improved using machine guns and other crew-served weapons mounted higher than the surrounding facility buildings. This allows for larger fields of fire over the tops of buildings. Communications are more effective when antennae have unobstructed surroundings.

Figure 3-3. Observation tower

Tactical Facility Construction

3-14. Observation towers should be constructed from a steel or thick wooden framework. A group of four towers with overlapping fires present an intimidating obstacle to any attacking force. As such, observation towers become prime targets of an attacking force. They must be designed and built to withstand attack from small-arms and RPG fire.

FIGHTING TRENCHES

3-15. Fighting trenches should be constructed approximately 5 feet high and 3 feet wide and include earthen parapets. These dimensions provide safe protection for a two-man litter carry. Trenches should be built in a zigzag pattern. Parapets should be built using a minimum of two layers of sandbags (capped with concrete) with multiple firing ports at varying intervals.

3-16. Full use should be made of all available natural building materials (such as trees, logs, and brush) for constructing and camouflaging covered and concealed emplacements and shelters. Manufactured materials (such as pickets, barbed wire, concrete [Figure 3-4], lumber, sandbags, and corrugated metal) are supplied by contractors or military support organizations. Captured enemy supplies, locally procured material, and demolished structures are additional sources of fortification building materials.

Figure 3-4. Indigenous employees mixing cement

3-17. When native materials are not available, units may use stackable barrier systems for vertical support and protection. A detailed list of construction materials and standards is provided in Table 3-1, page 3-6, and Table 3-2, page 3-7.

Table 3-1. Minimum thickness (in inches) required for protection against enemy fire

Material	7.62-mm Small Caliber and Machine Gun Fire at 100 Yards[1]	76-mm Antitank Rifle at 100 Yards	20-mm Antitank Fire at 200 Yards	37-mm Antitank Fire at 400 Yards	50-mm Antitank Fire at 400 Yards	75-mm Direct Fire at 500–1000 Yards
Solid Walls[2]						
Brick Masonry	18	24	30	60	–	–
Concrete (Not Reinforced)[3]	12	18	24	42	48	54
Concrete (Reinforced with Steel)	6	12	18	36	42	48
Stone Masonry	12	18	30	42	54	60
Timber	36	60	–	–	–	–
Wood	24	36	48	–	–	–
Walls of Loose Material Between Boards[2]						
Brick, Rubble	12	24	30	60	72	–
Clay (Dry)[4]	36	48	–	–	–	–
Gravel, Small Crushed Rock	12	24	30	60	72	–
Loam (Dry)[5]	24	36	48	-	-	–
Sand (Dry)[4]	12	24	30	60	72	–
Sandbags						
Brick, Rubble	20	30	30	60	70	–
Clay (Dry)[4]	40	60	-	-	-	–
Gravel, Small Crushed Rock	20	30	30	60	70	–
Loam (Dry)[5]	30	50	60	-	-	v
Sand (Dry)[4]	20	30	30	60	70	v
Parapets						
Clay[4]	42	60	–	–	–	–
Loam[5]	36	48	60	–	–	–
Sand[4]	24	36	48	–	–	–
Snow and Ice						
Frozen Snow	80	80	–	–	–	–
Frozen Soil	24	24	–	–	–	–
Icecrete (Ice and Aggregate)	18	18	–	–	–	–
Tamped Snow	72	72	–	–	–	–
Unpacked Snow	180	180	–	–	–	–

Notes:
[1] One burst of 5 rounds.
[2] Thicknesses to nearest 6 inches.
[3] Plain formed 3000 pound-per-square-inch (psi) concrete walls.
[4] Add 100% to thickness if wet.
[5] Add 50% to thickness if wet.
Except where indicated, protective thicknesses are for a single shot only. Where weapons place five or six direct-fire projectiles in the same area, the required protective thickness is approximately twice that indicated. Where no value is given, material is not recommended.

Table 3-2. Minimum thickness (in inches) required for protection against high-explosive shaped charges

Material	73-mm Recoilless Rifle	82-mm Recoilless Rifle	85-mm RPG-7	107-mm Recoilless Rifle	120-mm Sagger
Aluminum	36	24	30	36	36
Concrete	36	24	30	36	36
Granite	30	18	24	30	30
Rock	36	24	24	36	36
Snow (Packed)	156	156	156	–	–
Soil	100	66	78	96	96
Soil (Frozen)	50	33	39	48	48
Steel	24	14	18	24	24
Wood (Dry)	100	72	90	108	108
Wood (Green)	60	36	48	60	66
Notes: Thicknesses assume impact is perpendicular to surface. Protective thicknesses are for a single shot only. Where no value is given, material is not recommended.					

3-18. Materials and construction techniques used throughout Vietnam to construct SF "base camps" varied widely. The following vignette depicts the adaptability, ingenuity, and resourcefulness of the individual Vietnam-era SF Soldiers in their use of construction materials. This same resourcefulness can be found among present-day SF Soldiers operating in a wide range of environments.

Vietnam Rural Tactical Facility Construction Materials

Cement, even when available through normal military supply channels, was used sparingly in the isolated and remote rural TACFACs of Vietnam. The extreme weight and the enormous quantities required in TACFACs prevented large-scale distribution. A modest amount of commercial cement was available from Vietnam's Mekong Delta and distributed in 95-pound paper bags.

Although supplies of sand and gravel were uncommon, laterite, a suitable substitute for sand and gravel, was readily available. Laterite is normally found in tropical regions and used as a replacement for sand and gravel in many areas of Vietnam. It is a reddish mixture of clay, iron, and aluminum oxides and hydroxides formed by the weathering of basalt under steamy tropical conditions. Laterite, when used to make concrete, requires more cement than standard sand and gravel mixes; as such, there is a resulting loss in wear resistance. When the mix is sprinkled with water and turned over frequently, it could be compressed and made strong enough for such surfaces as roads, helicopter pads, and airfield runways, particularly when reinforced with steel rods, mesh, or engineer stakes.

During the Vietnam War, there was a shortage of cement, barbed wire, and precut dimensional lumber. Knowing that lumber decays quickly in tropical climates, the French used concrete in several innovative ways, including railroad ties, guard and sentinel posts, and vertical electric and utility poles. Moreover, when TACFACs were no longer needed, they were completely disassembled. All useful materiel was collected and moved to new sites or to other camps in need of material.

> Local Vietnamese merchants supplying dimensional lumber helped stimulate the local economy. Other Southeast Asian countries and the Philippines also provided lumber. Conventional military supply channels turned out a plethora of burlap and plastic sandbags. Additionally, items that were designed for one purpose evolved into another purpose—field-expedient construction material. Wooden or metal pallets, 55-gallon drums, metal shipping containers, corrugated sheeting and piping, cinder blocks, wire mesh, and chain-link fencing began to identify TACFACs throughout Vietnam.

CONCRETE

3-19. Concrete is very durable and an ideal building material for an SF TACFAC. It can be made even more durable by using steel reinforcing rods (rebar) when pouring concrete for an HLZ or airfield runway. Vibrating concrete increases the compressive strength and bond between concrete and rebar and decreases concrete permeability; decreases cold joints, honeycombing, excessive entrapped air and segregation; and causes concrete within a circular field of action to act like a liquid.

3-20. The SF unit may use a flex-shaft concrete vibrator, such as the Multiquip flex-shaft concrete vibrators designed to work in medium- to high-slump concrete. The Multiquip model BP25H is backpack-portable with a 2.1-horsepower Honda engine. Additional information regarding concrete-handling equipment may be found at www.multiquip.com/.

3-21. To use a concrete vibrator, the SF unit—

- Inserts the vibrator vertically, allowing it to penetrate rapidly to the bottom of the lift and at least 6 inches into the previous lift.
- Holds at bottom of lift for 5 to 15 seconds.
- Pulls the vibrator up at a rate of 15 seconds for a 4-foot lift (approximately 3 inches per second).

Note: To determine proper spacing when using a concrete vibrator, the SF unit should insert the vibrator so the fields of action overlap. The concrete must be observed to determine the vibrator field of action; high-powered vibrators and high-slump concrete have large fields of action. As a rule of thumb, the field of action is 8 times the diameter of the vibrator head.

3-22. The SF unit should stop vibrating the concrete when—

- The concrete surface takes on a sheen.
- Large air bubbles no longer escape.
- The vibrator changes pitch or tone.
- A change in vibrator action is felt.

Note: The vibrator should not be run outside of the concrete, as it will cause it to overheat. Vibrators should not be used to move concrete horizontally. Vibrators should not be forced or pushed into concrete, as it may not remain vertical and may get caught in the reinforcement.

ROCK

3-23. Direct and indirect fire fragmentation penetration into rock depends on the physical properties of the rock and the number of joints, fractures, and other irregularities contained therein. These irregularities weaken rock and can increase penetration. Several layers of irregularly-shaped rock can change the angle of penetration. Hard rock can cause a projectile or fragment to flatten and stop penetration.

BRICK AND MASONRY

3-24. Brick and masonry (Figure 3-5) provide excellent building materials, simplifying the precise construction of buildings and fighting positions. Direct and indirect fire fragmentation penetration into brick and masonry has the same protection limitations as rock.

Figure 3-5. Brick slab roofing

WOOD

3-25. Direct and indirect fire fragmentation protection using wood is limited because of its low density and relatively low compressive strengths. Wood is generally used as structural support for bunkers and buildings. Wood is less effective than soil for protection against penetration, and, with its low ignition point, it is easily destroyed by fire. In a tropical climate, wood is subject to weather extremes, animals, birds, insects, and harmful fungus that degrades wood in short order.

3-26. Building a roof using native hardwoods (at least 6 inches in diameter), the SF unit installs two layers of logs as a roof. The second log layer is installed perpendicular to the first, and both layers provide an overhang of 12 to 24 inches. Layers of sandbags and corrugated sheeting may be positioned between and above the log layers to help absorb mortar blasts. Because the level of protection afforded by these layers results in a sizeable increase in roof weight, more and larger vertical supports will be required.

SANDBAGS

3-27. The walls of buildings and fighting positions are built of sandbags in much the same way bricks are used; however, they should be used only to protect—not to support—the roof. When building with sandbags, a minimum depth of two layers is used. Sandbags also are useful for retaining-wall revetments. Most sandbags are made of an acrylic fabric that is rot and weather resistant. In most climates, plastic-based sandbags have a lifespan of 2 years with minimal deterioration. Older style burlap-based bags deteriorate much more quickly, particularly in tropical climates. The useful life of sandbags may be prolonged by filling them with a mixture of dry earth and cement, normally in the ratio of 1 part cement to 10 parts dry earth. The cement sets as the bags take on moisture. For sand and gravel mixtures, a 1 to 6 ratio is used. Filled bags may also be dipped and coated in cement-water slurry. Each sandbag is then pounded with a flat object, such as a section of lumber, to make retaining walls more stable.

3-28. Sandbags can be used for revetting walls or repairing trenches when the soil is very loose and requires a retaining wall. A sandbag revetment will not remain standing if it has a vertical face. The sandbag face must have a slope ratio of 1 to 4 toward the top of the wall—leaning against the bare earth it holds in place. The base for the sandbag revetment must stand on firm ground. To construct a sandbag revetment wall, the SF unit—

- Fills sandbags about three-quarters full with earth or a dry soil-cement mixture, ties the choke cords, and tucks the bottom corners of the bags after filling.
- Constructs the bottom row of the revetment by placing all bags as headers.
- Continues building the wall using alternating rows of stretchers and headers, with the joints broken between courses.
- Finishes the wall by making the top row of the revetment of headers.

3-29. Sandbags are positioned so that the planes between the layers have the same pitch as the base—at right angles to the slope of the revetment. All bags are placed so that side seams on stretchers and choked ends on headers are turned toward the revetted face. As the revetment is built, it is backfilled to shape the revetted face to the slope.

3-30. Often, the required amount of filled sandbags exceeds the capabilities of Soldiers and HN personnel equipped only with shovels. When the bags are filled from a large sand stockpile and using a lumber or steel funnel, the job is performed faster and more efficiently.

3-31. Sandbags are like any other piece of military equipment; once filled, they must be maintained, repaired, and replaced as necessary. Sandbags are similar in principle to the bulletproof vest—they are intended to contain, slow, dissipate, and stop the kinetic energy of high-speed projectiles and fragments. The constant onslaught of weather, insects, foot traffic, and enemy fire degrades the integrity and performance of sandbags.

PALLETS

3-32. Standard wooden cargo pallets measure 48 inches by 40 inches. Aluminum USAF 463L cargo pallets measure 108 inches by 88 inches. Both pallet types may be used as expedient construction material and may be integrated into revetted walls or flooring to provide extra support.

AMMUNITION CONTAINERS

3-33. Large wooden ammunition crates (such as those used to store artillery shells) and small ammunition boxes (such as those containing mortar rounds) are extremely useful and may be recycled to serve as flooring, shelving, ceilings, or field-expedient furniture. Ammunition containers also may be filled with dirt or sand and stacked to serve as makeshift sandbags. Because ammunition crates are not as stable as sandbags when stacked to any height, they must be supported by engineer stakes every few feet to prevent catastrophic collapse.

STEEL DRUMS

3-34. Empty steel drums can be used to make field-expedient solid waste burn-barrel latrines. To create a burn-barrel latrine, Soldiers cut a 55-gallon drum into halves. The barrel halves are partially filled with sand and placed underneath raised latrine seats. When required, the drum halves are pulled out, diesel fuel is added to the waste and lit, and the waste is allowed to burn in the open. Such burn barrels are cost-efficient and sanitary; however, they do require personnel, fuel, fire extinguishers, and shovels. Additionally, burning waste produces copious amounts of deep black smoke with a noxious odor.

3-35. Liquid waste processing requires the construction of a field-expedient urine tube. To create the urine tube, the SF unit first digs a hole large enough to hold a perforated 55-gallon drum. The top of the drum is removed, and a plastic tube—approximately 3 to 6 inches wide and 6 feet long—is inserted into the drum at a depth of 3 to 4 feet. The drum is then filled with crushed rock or gravel and topped off with a layer of soil or sand. The tube extends 2 to 3 feet above ground level and the open end is covered with mosquito wire mesh. The completed drum assembly acts as a small, environmentally friendly leach field.

3-36. Steel 55-gallon drums have a number of additional useful applications in a TACFAC. SF Soldiers may remove one end of the drum, fill it with dirt or sand, and integrate it into a stackable barrier system. If 463L pallets are unavailable for roofing (and steel drums are plentiful), both ends of a drum may be removed and it may be flattened out and used as sheeting for additional roof or structure support.

3-37. In addition to their structural applications, steel drums also may be integrated into the facility defense plan. Drums may be filled with jellied fuel, napalm, or suitable flammable liquid alternatives and then rigged with command-detonated explosives. These drums are positioned along avenues of approach in the outer barrier. When command-detonated, not only will the fireball impede enemy progress and undermine enemy morale, it will also provide a source of illumination for defending forces during hours of darkness.

LARGE SHIPPING CONTAINERS

3-38. Large, watertight CONEX-type shipping containers are used in a variety of roles, such as field-expedient sleeping quarters, arms rooms, ammunition bunkers, and food-storage sheds. When positioned aboveground, CONEXs should be lined with protective sandbag walls. Ideally, heavy construction equipment is available to excavate a hole and then position CONEXs belowground.

METAL PIPE

3-39. Corrugated metal piping is made in round or half-round sections ranging from a few inches to 8 feet in diameter. Most corrugated metal piping is zinc coated to protect against rust. Half-round and flanged pipe in diameters of 1 to 2 feet may be used for drainage and other purposes. Full-round corrugated pipes are used as culverts. Half-round sections of corrugated piping are stronger than the 463L metal pallets. These sections may be used as field-expedient bunker roofs, floors, and walls. Because corrugated metal pipes offer little protection against direct or indirect fire, sandbags must be stacked against any piping used as protection for U.S. personnel or equipment.

CINDER BLOCKS

3-40. The standard-size U.S. cinder block is 16 by 8 by 8 inches. Cinder blocks are particularly useful in tropical climates, where sandbags are prone to fail. These blocks may be used to construct walls of buildings and fighting positions, as well as to elevate personnel shelters, bunkers, and other buildings above flood level. When the dual cavities of cinder blocks are packed with concrete, sand, or dirt, they provide additional protection against projectiles.

3-41. Soldiers in the Vietnam era used construction resources similar to those being used in Iraq today. The key difference, however, is that there are greater numbers of local and international contractors available in the Iraq theater. Also, there is outstanding military and contract air and ground resupply presently available. The following vignette demonstrates the resourcefulness and adaptability of SF Soldiers in Vietnam. Their adaptability and resourcefulness is alive and well in Iraq, the Philippines, and Afghanistan today.

> **Vietnam Rural Tactical Facility Construction Resources**
>
> Given the remote locations where rural TACFACs were positioned in Vietnam, a variety of joint and combined support units, as well as nongovernmental organizations (NGOs), assisted in their construction. Conventional engineer units were attached to SF units to provide specialized, technical support, such as well-drilling teams that could drill down more than a quarter of a mile. Most technically complex water and electric work was contracted out to Vietnamese or Philippine civilian engineers.
>
> The 5th Special Forces Group (Airborne) (SFG [A]) had engineer personnel on staff, and the senior engineer had an electric utility unit and staff engineers that were assigned to SF companies. These personnel built, upgraded, maintained, and rebuilt over 150 SF TACFACs between 1962 and 1971. Attached CA advisory squads had heavy equipment and remained on site long enough to train the local civilian irregular defense group (CIDG) personnel on construction techniques. Working in a train-the-trainer capacity, the engineers and advisory squads built one of each type structure or position. Local HN personnel would then complete the majority of the project. Working on rotating schedules, one group would work while the other group conducted patrols and provided security.
>
> United States Navy (USN) Seabees provided well-drilling capability and a variety of heavy equipment in Vietnam. They also introduced the Cinva-Ram—a simple, manually operated machine that makes concrete blocks. Developed in Colombia during the 1950s, the Cinva-Ram used local soil for block making in remote and rural locations. A 6-foot lever on the machine compressed concrete with 2,000 pounds per square inch. Additional information on the Cinva-Ram and the manufacture of compressed earth blocks may be found at:
>
> http://kubuildingtech.org/ngore/nilsweb/cinvablocks/index.html
>
> U.S. Army combat engineers did not build many SF TACFACs in Vietnam; however, when tasked to do so, the facility usually was finished ahead of schedule. Even with such skilled engineer support, it took between 60 and 150 days to complete a permanent SF TACFAC. The indigenous CIDG were crucial to the TACFAC effort by providing the essential manual labor.

FENCING

3-42. Chain-link and wire-mesh fencing may be used as field-expedient shielding against incoming RPGs or other shape-charged projectiles fired at protected buildings or bunkers. Innovative anti-RPG wire-mesh barriers also may be fabricated for vehicles, large diesel-electric generators, and fuel and ammunition storage sites.

EMERGENCY SIGNAL PROJECT

3-43. The SF unit may construct a fire arrow to serve as an emergency signal. To construct a fire arrow, the SF unit makes an arrow shaft (12 feet by 1 foot) and two arrow-point sections (4 feet by 1 foot). An empty 4-foot wooden cable reel is used as a rotating arrow base that can quickly and easily lock into the proper azimuth. The SF unit ensures that redundant signaling tools (such as flares, fire pots [mixing diesel fuel and sand], and strobes) are prepared and available for use.

3-44. When using a fire arrow as an emergency signal during daylight hours, the SFOD uses smoke pots, smoke grenades, strobe lights, flares, VS-17 panels, and eye-safe lasers. At night, fire pots, flares, strobe lights, chemical lights, glint tape, lasers, and infrared emitters may also be used.

Chapter 4
Tactical Facility Operations and Defense

A TACFAC is established to allow the SFOD to temporarily defend itself in preparation for an offense operation. The base-defense plan is how an SFOD intends to secure the TACFAC in the event of an enemy assault on the TACFAC. The base-defense plan consists of coordinating barrier systems with direct and indirect fires. Established responsibilities thoroughly coordinate protection, in a given operational environment, with a comprehensive fire support plan that allows the execution of the base defense. In order to ensure survivability, the base-defense plan may require indirect, close combat attack (CCA), and close air support (CAS) fires to augment the TACFAC organic direct and indirect fire assets. As such, SFODs must plan and coordinate joint fires into the base-defense plan.

AREAS OF RESPONSIBILITY

4-1. The base commandant is responsible for establishing the long-term vision and priorities-of-work for the TACFAC. The commandant must consider available resources (such as manpower, building materials, and time available) and weigh them against the operational needs of the facility (such as C2, training requirements, and sustainment requirements). The base commandant ensures redundant systems by establishing contingency plans when possible, and periodically reviews the progress and adjusts the plan as needed. The commandant delegates responsibility of functions as needed to allow proper maturity of the TACFAC.

4-2. The facility manager controls the construction and maintenance of the facility through its development. Translating the intent of the base commandant and applying the resources needed to accomplish that intent, the facility manager supervises the labor force to ensure time and energy is applied in the most efficient means possible. The facility manager is also responsible for the usage of facilities (such as training areas or meeting areas) to deconflict scheduling.

4-3. The security manager ensures that the facility is protected from all possible threats and that appropriate risk management is applied in all areas in a given operational environment. The security manager establishes systems that restrict access to areas within the TACFAC and ensures that construction requirements (such as bunkers or fighting positions) are given at the appropriate level of priority by the facility manager. The security manager must establish systems that maintain operations security (OPSEC) and vet anyone granted access into the facility.

4-4. The medical coordinator ensures appropriate consideration is given to such medical impacts as location of SWEAT-MS facilities. The medical coordinator advises the base commandant of the potential medical effects in the planning and execution phases and establishes systems to handle medical emergencies.

OPERATIONAL ENVIRONMENTS

4-5. The SFOD plans and conducts SO in permissive, hostile, or uncertain operational environments. Understanding the operational environment and the threat potential are critical to the execution of security, protection, and TACFAC defense. Operational environments are composites of the conditions, circumstances, and influences that affect the employment of capabilities and bear on the decisions of the commander. There are as many operational environments as there are commanders and specific operations.

Permissive Environment

4-6. A permissive environment is an operational environment in which HN military and law enforcement agencies (LEAs) have control and the intent and capability to assist operations that a unit intends to conduct. A permissive operational environment is the most preferable environment for a TACFAC because enemy offensive action is not likely. TACFAC defense measures may be as passive as increasing counterintelligence (CI) sweeps of local HN areas, coordinating with the regional security officer in the embassy, or erecting chain-link fences around vital locations.

Hostile Environment

4-7. A hostile environment is an operational environment in which hostile forces have control and the intent and capability to effectively oppose or react to the operations a unit intends to conduct. Hostile operational environments are the least preferred, most resource-intensive, and most dangerous of the three types of operational environments. Enemy intentions, capabilities, and potential are known, and friendly units must take necessary actions to address the threat.

Uncertain Environment

4-8. An uncertain environment is an operational environment in which HN forces, whether opposed to or receptive to operations that a unit intends to conduct, do not have totally effective control of the territory and population in the intended operational area. An uncertain operational environment has characteristics of both permissive and hostile operational environments. TACFAC defense is more overt and active in uncertain environments. Operations in such environments require more training in the ROE.

THREAT LEVEL

4-9. Military Police (MP) or Infantry units should be used to aid in the defense of forward-deployed bases. FM 3-19.1, *Military Police Operations*, states that base commanders are responsible for addressing all Level I threats to the base. Although a TACFAC commander may request MP or Infantry augmentation at any threat level, the overall TACFAC defense always remains the responsibility of the TACFAC commander. In accordance with (IAW) FM 3-19.1, threats are divided into three levels. These levels provide a general description and categorization of threat activities, identify the defense requirements to counter them, and establish a common reference for planning guidelines.

Level I

4-10. Level I threats are the lowest of the three threat levels; however, Level I threat environments are not considered "safe" because of the enemy's ability to operate clandestinely. It is very resource intensive when the complexity of finding and defeating an enemy presents few (if any) overt indicators. Level I threats include the following types of individuals or activities:

- *Enemy-controlled agents.* Enemy-controlled agents are a potential threat to the SF TACFAC. These elements may include the clandestine and covert elements of the enemy's forces, such as special-purpose forces, military intelligence (MI) personnel, and CI forces. Their primary missions include espionage, sabotage, subversion, and criminal activities. Their activities span the range of military operations and may include assassinating or kidnapping key military or civilian personnel or guiding special-purpose individuals or teams to the SF TACFAC.
- *Enemy sympathizers.* Civilians sympathetic to the enemy may become significant threats to the SF TACFAC. They may be the most difficult to neutralize because they normally are not part of an established enemy-agent network, and their actions are random and unpredictable. Indigenous groups sympathetic to the enemy or those simply opposed to the United States can be expected to provide assistance, information, and shelter to guerrilla and enemy unconventional or special-purpose forces operating in the area.
- *Civil disturbances.* Civil disturbances (such as demonstrations and riots) involving local issues and not directly involving enemy sympathizers may pose a direct or indirect threat to SF TACFACs.

LEVEL II

4-11. Level II threats are based on the enemy's potential to conduct operations against the SF TACFAC. Level II threats represent more than just intelligence-gathering efforts. Examples of Level II threats include the following:

- *Regular forces.* Small tactical units, such as specially organized reconnaissance elements, are capable of conducting raids and ambushes in addition to their primary reconnaissance and intelligence-gathering missions.
- *Irregular forces.* Resistance and insurgent forces pose a serious threat to the SF TACFAC and the orderly conduct of the local government and services. They may establish and activate espionage networks, collect intelligence, carry out specific sabotage missions, develop target lists, and conduct damage assessments.
- *Terrorists.* Terrorists are among the most difficult threats to neutralize and destroy. Their actions span the full range of military operations.

LEVEL III

4-12. Level III threats are the highest threat level, and may have disastrous effects on the ability of U.S. forces to successfully execute strategic operations. Level III threats are made up of conventional forces capable of projecting combat power rapidly by land, air, or sea. Specific examples include airborne, heliborne, and amphibious operations; large, combined, ground-force operations; and infiltration operations involving large numbers of individuals or small groups infiltrated into friendly-controlled areas, regrouped at predetermined times and locations, and committed against priority targets. Level III forces may use a combination of the following tactics as a precursor to a full-scale offensive operation:

- *Air or missile attack.* Threat forces may be capable of launching an air or missile attack against the SF TACFAC. It is often difficult to distinguish quickly between a limited or full-scale attack before impact; therefore, protective measures should normally be based on the maximum threat capability.
- *Chemical, biological, radiological, and nuclear (CBRN) attack.* Commanders must be aware that CBRN munitions may be used in conjunction with air, missile, or other conventional force attacks. CBRN weapons may also be used at Level I or Level II by terrorists or unconventional forces in order to accomplish their political or military objectives.

JOINT FIRES

4-13. Joint fires are defined as fires delivered during the employment of forces from two or more components (Army, USAF, USN, or United States Marine Corps [USMC]) in coordinated action to produce desired effects in support of a common objective. Joint fires are weapons effects from joint operations and are provided to assist SO land, amphibious, and air forces; joint air operations; joint maneuver operations; and joint interdiction operations.

4-14. Joint fire support is defined as joint fires that assist air, land, maritime, and special operations forces (SOF) to move, maneuver, and control territory, populations, airspace, and key waters. Joint fire support may include, but is not limited to, the lethal effects of CAS) or CCA, artillery, rockets, missiles, and mortars, as well as the nonlethal effects of CA, IO, and PSYOP. The ability of the SFOD to synchronize joint fire support with security is essential during defensive operations.

4-15. Combining joint fires and security measures provides the SFOD needed firepower at decisive points to achieve surprise, psychological shock, momentum, and effects (death and destruction). These, in turn, assist the SFOD by not only destroying the enemy but also in shattering enemy morale and unit cohesion.

4-16. Face-to-face coordination is the preferred method for coordinating joint fire support with any unit providing support. Problems such as radio or frequency compatibility, language, call signs, authentication, safety considerations for munitions, and call-for-fire procedures should be resolved at these meetings.

4-17. Each SFG(A) now has a joint fires element (JFE) assigned. The JFE is made up of one major, one targeting warrant officer, and one senior noncommissioned officer (NCO), all from the Field Artillery (FA) branch. At the group level, the JFE advises the group commander and staff on the planning, synchronization, and execution of lethal and nonlethal fire support for subordinate elements. Additionally, each SF battalion also has a JFE assigned; this JFE is made up of one FA captain and one senior FA NCO. These personnel can provide in-depth knowledge to the SFODs on the full spectrum of fire support assets available to support TACFAC base defense.

INDIRECT FIRE

4-18. Artillery and mortars are reliable, 24-hour, all-weather fire support systems. Howitzers and mortars can provide fires with a variety of munitions within their respective ranges to support SFODs in their mission. Artillery can, and frequently is, attached to the SFOD at a TACFAC. In this case, the TACFAC serves as a firebase, and normally is referred to as a firebase.

4-19. Because of their limited firepower, SFODs must consider all available sources of fire support. SFODs usually have limited 60- or 81-millimeter mortars as organic fire support, but the limited range and the number of personnel required to man these systems may render them impractical during SO. SFODs may have access to other indirect fire assets in their AO based on other friendly units. These may include 120-millimeter mortars, 105- and 155-millimeter howitzers, high-mobility artillery rocket systems (HIMARSs), or multinational fire support systems. SFODs and fire support planners must consider the benefits and risks of these systems during planning. The location of units throughout the AO will determine what fire support is available. The distances involved or the geometry of the battlefield and the location and actions of the enemy will determine what assets can be used.

4-20. When the SFOD has priority of fires from artillery or mortars, the following indirect fire questions should be addressed:

- Where should the howitzer or mortar be positioned in order to support the operation?
- Will alternate or supplemental positions be needed?
- Is sufficient ammunition on hand to support the SFOD if in an extended contact scenario?
- Are the target reference points (TRPs) easily identifiable?
- Is there dead space that the howitzers cannot cover (due to the angle of fire) and can this space be covered with mortars instead?
- Are all members of the mission familiar with call-for-fire procedures?
- Are fires planned throughout the depth of the likely defensive engagement area?
- Are all obstacles covered by indirect fire?
- Have final protective fires been registered?
- Can specialty munitions (such as smoke and illumination rounds) enhance the defense?

CLOSE COMBAT ATTACK

4-21. U.S. Army CCA is defined as a coordinated attack by Army aircraft against targets that are in close proximity to friendly forces. During CCA, the attack team engages enemy units with direct fires that impact near friendly forces. The distance between targets and friendly forces may range from tens of meters to several thousand meters. CCA is coordinated and directed by a team, platoon, or company-level ground unit using the standard CCA brief (Figure 4-1, page 4-5). Once the aircrews receive the brief from the ground commander, they develop a plan to engage the enemy force while maintaining freedom to maneuver. Due to capabilities of the aircraft and the enhanced situational awareness of the aircrews, terminal control from ground units or controllers is not necessary. CCA is not synonymous with CAS.

1. Observer/Warning Order: " _____ , THIS IS _____ ; FIRE MISSION; OVER."   (Aircraft)   (Observer)
2. Friendly Location/Mark: "MY POSITION _____ , MARKED BY _____ ."   (TRP, Grid)   (Strobe, Beacon, Infrared [IR] Strobe)
3. Target Location: " _____ ."   (Bearing [magnetic] & Range [meters], TRP, Grid)
4. Target Description/Mark: " _____ , MARKED BY _____ ; OVER." (Target Description)   (IR pointer, Tracer)
5. Remarks: " _____ ."   (Threats, Danger-Close Clearance, Restrictions, At My Command)
As Required: 1. Clearance: Transmission of the CCA brief is clearance to fire (unless danger-close). Danger-close ranges are IAW FM 3-09.32, *Multiservice Procedures for the Joint Application of Firepower*. For closer fire, the observer/commander must accept responsibility for increased risk. State CLEARED DANGER CLOSE on line 5. This clearance may be preplanned. 2. At my command: For positive control of the gunship, state AT MY COMMAND on line 5. The gunship will call READY FOR FIRE when ready.

Figure 4-1. Close combat attack checklist

CLOSE AIR SUPPORT

4-22. According to JP 3-09.3, *Joint Tactics, Techniques, and Procedures for Close Air Support*, CAS is air action by fixed- and rotary-wing aircraft against hostile targets that are in close proximity to friendly forces and which require detailed integration of each air mission with the fire and movement of those forces. The primary difference between CAS and CCA is control. CAS requires a qualified joint terminal attack controller (JTAC) to control weapons release. A JTAC is a qualified (certified) service member who, from a forward position, directs the action of combat aircraft engaged in CAS and other air operations.

RADARS

4-23. The U.S. Army currently uses a variety of manned and automated countermortar and counterfire radars to enhance protection. These systems are either active or passive. Active systems use directed energy (microwaves) and track reflected energy. Active systems track incoming rounds at different ranges, providing both a predicted point of impact (POI) and a point of origin (POO). Passive systems use acoustic sounding from several microphones and compare the differences to determine a location or direction to where the round originated.

UNATTENDED TRANSIENT ACOUSTIC MEASUREMENT AND SIGNATURE INTELLIGENCE SYSTEM

4-24. The unattended transient acoustic measurement and signature intelligence system (UTAMS) (Figure 4-2, page 4-6) is a ground-based acoustic sensor array that provides accurate detection, geolocation, and classification of direct and indirect fires at standoff distances. Each of the UTAMS acoustic sensor arrays

Chapter 4

independently processes the detected events based on statistics from the signal content against the background noise, computes the line of bearing to the firing locations, and sends the line-of-bearing information to a central base-station laptop computer via a radio link. The base station performs source localizations via correlation and triangulation techniques. The approximate planning range of UTAMS is 13 kilometers.

Figure 4-2. Unattended transient acoustic measurement and signature intelligence system

LIGHTWEIGHT COUNTERMORTAR RADAR

4-25. The lightweight countermortar radar (LCMR) (Figure 4-3, page 4-7) detects and locates mortar firing positions automatically by detecting and tracking the mortar shell and then backtracking to the weapon position. The LCMR provides continuous 360-degree surveillance and mortar location out to a range of 7 kilometers. The LCMR system is designed for airborne operations and can be deployed in a door bundle. When a mortar is detected, the LCMR sends a warning message indicating an incoming round. Once sufficient data is collected, the weapons location is transmitted.

AN/TPQ-36 RADAR

4-26. The AN/TPQ-36 (Figure 4-4, page 4-7) is optimized to locate short-range, high-angle, low-velocity indirect fire weapons such as mortars and short-range artillery. It also can locate longer-range artillery and rockets within its maximum range. Planning ranges for the AN/TPQ-36 are 14.5 kilometers for artillery, 18 kilometers for mortars, and 24 kilometers for rockets. The minimum range of the system is 750 meters. These planning ranges are where the highest probability of detection lies for the system design. The AN/TPQ-36 normally has a three-man team that operates and maintains the radar.

Figure 4-3. Lightweight countermortar radar

Figure 4-4. AN/TPQ-36 radar

Chapter 4

AN/TPQ-37 Radar

4-27. The AN/TPQ-37 (Figure 4-5) is optimized to locate long-range, low-angle, high-velocity weapons such as long-range artillery and rockets. However, it will also locate shorter-range, higher-angle, lower-velocity weapons, thereby complementing the AN/TPQ-36. The planning ranges used as a baseline to position the AN/TPQ-37 are 30 kilometers for mortars and 50 kilometers for rockets. The minimum range is 3 kilometers. These planning ranges are where the highest probability of detection lies for the system design. The AN/TPQ-37 normally has a seven-man team that operates and maintains the radar.

Figure 4-5. AN/TPQ-37 radar

UNMANNED AIRCRAFT SYSTEMS

4-28. Unmanned aircraft system (UAS) (Figure 4-6) operations support battlefield commanders and their staffs as they plan, coordinate, and execute operations. UASs increase the situational awareness (SA) of commanders through intelligence, surveillance, and reconnaissance (ISR). Armed UASs provide commanders direct-fire capabilities to prosecute the close fight and influence shaping of the battlefield. Army UASs can perform some or all of the following functions: enhanced targeting through acquisition; detection, designation, suppression, and destruction of enemy targets; and battle damage assessment (BDA). Other UAS missions support the maneuver commander by contributing to the effective tactical operations of smaller units.

Figure 4-6. MQ-1 Predator unmanned aircraft system

MULTINATIONAL FIRE SUPPORT

4-29. Current operational environments often have SF elements working in a multinational-force AO. Although this often places SFODAs outside of the range of most U.S. fire support systems, the fire support assets of multinational partners may be used to support SF operations provided the appropriate level of coordination takes place. Support from multinational forces can range from providing supporting radar coverage at a firebase to placing howitzers in direct support of U.S. forces.

4-30. When foreign fire support assets are placed in support of SF, additional planning considerations and steps must be taken to ensure unhindered execution. Some items that must be taken into account and resolved prior to execution include the following:
- *Fire support requests.* What is the request and approval chain? Can the SFODA talk directly with the asset or do requests need to go through a higher HQ? How long does the approval process take?
- *Communications.* Does the supporting unit have communications systems that are compatible with U.S. systems? Will requests, corrections, and adjustments be given directly to the supporting unit or will a relay be needed? Are there communications security (COMSEC) issues? Is there a language barrier which must be overcome?

- *Tactics and techniques.* Is the supporting unit familiar with the U.S. call-for-fire format or does the supporting unit use a different system? Which format will be used? Are U.S. personnel familiar with the supporting unit call-for-fire format? Are the fire-direction procedures familiar or are instructions needed?
- *ROE.* What ROE do the multinational force follow? What are the differences between the ROE covering SF personnel in theater versus the multinational force? Do the multinational force ROE limit their ability to support U.S. forces in contact? What are the specific national caveats, if any? How do those caveats affect the mission?
- *Indirect assets.* What are the capabilities and limitations of the multinational fire support system? What munitions are available and what are their planning ranges? Will they need ammunition resupply and who is responsible for that resupply?
- *Fire support coordination measures (FSCMs).* Who will submit and disseminate appropriate FSCMs? Are both the multinational force and SF personnel familiar with FSCMs?

FIRE SUPPORT PLANNING

4-31. The SFOD plans for the employment of fire support (FS) assets such as indirect fires, CAS, and CCA throughout the AO to cover dead space from direct fires, to secure LOCs, and to establish TRPs for timely support. Commanders consider the requirements for FS assets throughout their AO based on their scheme of maneuver; however, if possible, the execution of the defense should not be tied to the availability of assets. Timing should also be considered when planning FS asset employment.

4-32. In order to provide timely fires, the supporting unit should maintain the appropriate munitions and, when possible, rehearse the mission to enhance SA. All of these measures are conducted to reduce the response time for fires to the supported unit. CAS and CCA planning must consider the depth and breadth of the entire battlefield. As part of the planning process, FSCMs must be established to prevent catastrophic results from hasty employment and to ensure proper deconfliction at the appropriate levels for lethal fire support. Both lethal and nonlethal fire support must be considered during mission planning. Sometimes the use of a PSYOP broadcast, a medical civic action program (MEDCAP) mission, or a leaflet drop—all considered nonlethal fire support—can give the SFOD the desired results without ever firing a shot.

UNMANNED AIRCRAFT SYSTEM PLANNING CONSIDERATIONS

4-33. The capabilities of UASs expand the planning time and increase SA; however, there is the potential for over-reliance on UASs at the expense of other collection assets. This can result in an ISR collection plan that is neither comprehensive nor integrated. All intelligence-gathering assets require equal consideration and emphasis when developing a focused ISR plan that meets the CCIRs.

4-34. Many UASs supporting SF Soldiers are armed and can be used to conduct strikes in support of the mission. When using a UAS platform in an offensive role, additional steps must be taken to ensure the safety of forces on the ground and to make certain the proper target is engaged. The requirements for conducting armed UAS missions are the same as CAS; the ground-force commander must authorize the strike. A current and qualified JTAC must provide the 9-line report (Figure 4-7, page 4-11), conduct target talk-on, and provide terminal control. Positive identification (PID) must be obtained and the appropriate ROE must be followed. Each theater has specific requirements for the strike requests and approval process. Consultation with the unit Judge Advocate General (JAG) is recommended when conducting fires on targets.

DEFENSIVE EMPLOYMENT OF FIRE SUPPORT

4-35. Fire support in defensive actions is characterized by centralized planning with centralized execution. Indirect-fire targets should be planned in depth throughout the AO. Any obstacles in the defensive battlefield should have preplanned targets assigned to them. Registration of howitzers and mortars and the adjustment of final protective fires (FPFs) should be completed as time allows.

> 1. Initial point (IP). A known position on the ground.
> 2. Heading from the IP to the target.
> 3. Distance from the IP to the target in nautical miles.
> 4. Target elevation in feet above mean sea level.
> 5. Target description.
> 6. Target location coordinates.
> 7. Type of mark, smoke, laser, etc.
> 8. Location of friendly forces from the target, cardinal direction, and distance in meters.
> 9. Egress direction and/or control point after attack.

Figure 4-7. Standard 9-line report format

4-36. Although FS assets are inherently offensive, FS also can be effectively employed to support the defense. In defensive operations, FS assets can be planned and used to cause the enemy to deploy prematurely, to deny the enemy access to critical terrain, or to slow or stop the enemy's attack against facilities, checkpoints, or mission support sites (MSSs). FS assets can be assigned to support both the mobile and fixed elements of the defense. The SFOD may plan the defensive use of FS assets to—

- Support the TACFAC fire support plan.
- Support the maneuver elements or quick reaction forces (QRFs) moving to aid the element in contact.
- Attack enemy forces that have penetrated friendly positions.

SECURITY OF LINES OF COMMUNICATIONS

4-37. The SF TACFAC may be provided with additional firepower to keep LOCs open and supplies and equipment moving throughout the SFOD area. CAS and CCA aircraft flying the planned routes prior to movement of QRFs may detect enemy personnel lying in ambush or emplacing IEDs. Preplanned TRPs on likely threat locations along QRF routes to the TACFAC will reduce fire-mission processing time.

CLOSE AIR SUPPORT AND CLOSE COMBAT ATTACK REQUESTS

4-38. There are two types of CAS and CCA requests available to the SFOD—preplanned and immediate. Preplanned requests may be filled with either scheduled or on-call air missions, whereas immediate requests are filled with on-call missions or diverted, previously scheduled aircraft.

Preplanned Requests

4-39. Preplanned requests are CAS or CCA requirements the SFOD has foreseen early enough during mission planning to be included in the joint air tasking order (ATO) cycle. Defensive plans inherently cannot be preplanned; however, TRPs can be coordinated throughout the AO. Because preplanned CAS must be controlled by a JTAC, the SFOD may be required to request JTAC support through its AOB and SOTF as soon as the requirements for CAS are identified. The SFOD must submit a preplanned request for CAS and CCA prior to the requested cutoff time. Submission procedures for preplanned requests (such as the numbering system and the cutoff time for inclusion in the ATO) are theater-specific, and detailed guidance should be found in the combined joint special operations task force (CJSOTF) SOPs. CAS and CCA planners prepare requests using Department of Defense Form (DD Form) 1972 (Joint Tactical Air Strike Request). JP 3-09.3 provides detailed instructions for completing DD Form 1972.

Immediate Requests

4-40. Immediate requests arise from situations that develop outside the ATO planning cycle, such as time-sensitive targets or troops in contact. Because these requirements cannot be identified early on, tailored ordnance loads may not be available for specified targets. Although support for the defense cannot be forecasted, it can be planned and coordinated. Established TRPs can reduce the time required for the supporting unit to provide fires for the TACFAC. Rehearsed defensive plans by CAS and CCA assets can enhance SA and reduce talk-on time.

Troops in Contact

4-41. When an SFOD has troops in contact, the JTAC will coordinate directly with the SOTF and air support operations center (ASOC) for immediate CAS or CCA. CAS or CCA for troops in contact will normally be given priority over all other CAS and CCA missions and may be executed with SFOD members who are not JTAC-qualified (Figure 4-8).

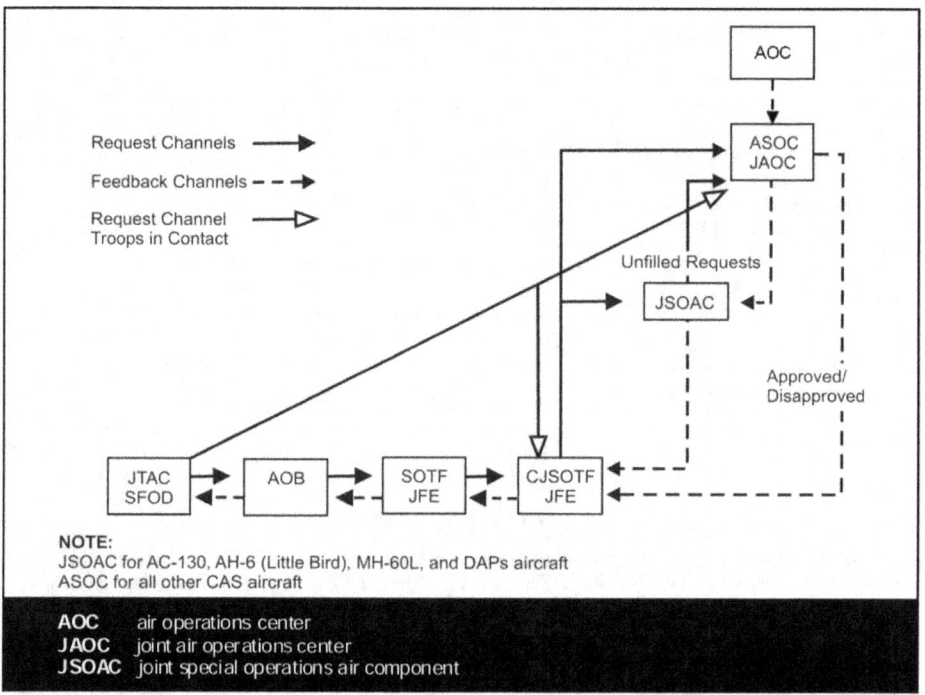

Figure 4-8. Special Forces immediate close air support request channels

4-42. SFOD personnel can effectively call CAS during contact situations, but a JTAC is normally required. In circumstances where the SFOD requires CAS and no JTAC-qualified personnel are available, the SFOD commander must consider the risk and accept full responsibility for the results of the attack. The non-JTAC controller must—

- Indicate to the aircraft (when the aircraft checks in) that he is not JTAC qualified.
- Provide as much of the 9-line report as possible.
- Pass target elevation, target location, and restrictions (at a minimum).

4-43. In return, when dealing with non-JTAC controllers, aircrews will—

- Assist the SFOD to the greatest extent possible in order to bring fires to bear.
- Pull information needed to complete the 9-line briefing.
- Exercise vigilance with target identification, weapons effects, and friendly locations.

Tactical Facility Operations and Defense

BASIC CLOSE AIR SUPPORT AND CLOSE COMBAT ATTACK PLANNING

4-44. Basic CAS and CCA planning begins with an analysis of METT-TC. The JFE is included in the beginning of planning to ensure the request for CAS arrives prior to the closure of the ATO cycle or the CCA request is submitted prior to local deadlines. Additional factors will determine the tactics and techniques required to conduct a particular CAS/CCA mission.

4-45. In addition to U.S. aircraft on the daily ATO, multinational forces also contribute aircraft. Although all CAS-designated aircraft support ground forces, each nation may have specific national caveats. These caveats may limit or restrict specific aircraft from certain countries. These caveats are found in the special instructions (SPINS) that are published with the ATO. Commanders, staff planners, and JFEs must be familiar with these caveats and their potential impact on the mission.

Mission

4-46. The SFOD plans the base defense, including the specified and implied tasks to be performed in accomplishing the objective, and the commander's intent. Understanding the key purpose of the mission allows supported and supporting units the latitude to exercise initiative and exploit opportunities. In planning CAS and CCA missions, the JFE should understand the ground commander's intent, scheme of maneuver, C2 requirements, civil aspects of the AO, and criteria for specific ROE. This understanding increases overall situational understanding by all participants and facilitates the initiative required to maximize CAS and CCA effectiveness.

Enemy

4-47. By determining key enemy characteristics, such as composition, disposition, order of battle (OB), capabilities, and likely COAs, the SFOD can begin to formulate how CAS and CCA can best be integrated. Other considerations include enemy C2 capabilities and the potential or confirmed presence of chemical or biological contamination. From this information the JTAC must anticipate the enemy ability to affect the mission and the potential influence of enemy actions on flight tactics and techniques. The potential for the enemy situation to change during the course of the mission emphasizes the importance of communications and close coordination between the aircrews and the JTAC or controller. In-flight updates on enemy activity and disposition along the flight route and in the TACFAC area may require the aircrews to alter their original plan and tactics. If the enemy is successful at disrupting communications, alternatives are planned to ensure mission accomplishment. Secure voice equipment and frequency-agile radios can overcome some effects of enemy interference.

Terrain and Weather

4-48. A terrain survey, developed by the JTAC, is used to determine the best routes to and from the TACFAC area. Where the terrain permits and when the threat dictates, flight routes should maximize the use of terrain masking to increase survivability against air defense systems. When practical, flight routes, holding areas, IPs, release points (RPs), and battle positions (BPs) should use terrain features that are easily recognizable—both day and night. Broad-area satellite imagery and air mission planning and rehearsal systems can assist in selecting optimum flight parameters.

4-49. Weather plays a significant role in CAS and CCA operations. It influences both enemy and friendly capabilities to locate, identify, and accurately attack CAS and CCA targets. Weather can also influence the effectiveness of laser designators, precision-guided munitions (PGMs), night vision devices (NVDs), and thermal imaging systems. Planners at every level require an understanding of the effects that weather can have on CAS and CCA aircraft navigation, sensors, and weapons systems.

Troops and Support Available

4-50. Commanders must have a thorough knowledge of friendly force troop locations in the AO. CAS and CCA operations must be fully integrated into the supported commander's scheme of maneuver and the fire support plan. Under normal planned CAS scenarios, a certified JTAC is required. When an SFOD has troops in contact, the JTAC will coordinate directly with the SOTF and ASOC for immediate CAS or

Chapter 4

CCA. CAS and CCA for troops in contact are normally given priority over all other CAS or CCA missions and may be executed with SFOD members who are not JTAC-qualified. Available aircraft platforms may have limitations that restrict their ability to fully support a defense. Restrictive terrain may limit CAS platforms, and high altitude may restrict CCA platforms.

Time Available

4-51. Time is the critical element in coordinating events and massing fires to achieve the desired effects on the ground. All possible measures must be taken early to increase available time during execution of the defense. Fire support planners must estimate the amount of time necessary for the aircraft to execute the mission. Inadequate time management may result in reduced effectiveness and increased risk to aircrews and troops on the ground.

Civil Considerations

4-52. The impact of combat operations on noncombatants and civilian structures must be taken into account. Collateral damage estimates (CDEs) and proper weaponeering of targets must be considered during both the planning and execution of fire support operations.

AIR SUPPORT BASIC CONDITIONS

4-53. There are nine basic conditions that optimize the execution of CAS and CCA operations. Although these conditions are not absolutely required to conduct CAS and CCA operations, creating the environment and shaping the battlefield to favor these conditions increases the probability of success. The nine basic conditions are—

- Air superiority.
- Suppression of enemy air defenses (SEAD).
- Target marking and friendly marking.
- Favorable weather.
- Prompt response.
- JTAC and SFOD skill.
- Weaponeering.
- Communications and information systems.
- C2.

Air Superiority

4-54. Air superiority permits CAS and CCA operations to function more freely and denies the same advantage to the enemy. Air superiority may range from local or temporary control of the air to control over the entire theater. JP 1-02, *Department of Defense Dictionary of Military and Associated Terms*, defines air superiority as "that degree of dominance in the air battle of one force over another that permits the conduct of operations by the former and its related land, sea, and air forces at a given time and place without prohibitive interference by the opposing force." This involves negating enemy airborne and ground systems, including air-to-air, air-to-surface, surface-to-air, and electronic warfare (EW) systems to a level at which they are incapable of having an adverse effect on friendly force operations.

Suppression of Enemy Air Defenses

4-55. SEAD may be required for CAS and CCA aircraft to operate within areas defended by enemy air defense systems. Available methods to suppress enemy air defense threats include destruction and disruption.

Target Marking and Friendly Marking

4-56. The requesting commander can improve CAS and CCA effectiveness by providing timely and accurate target marks. Target marking aids CAS and CCA aircrews in building situational understanding and in locating and attacking the proper target. Pointers, lasers, and colored smoke can all assist in marking targets for CAS and CCA. SFODs also must consider air-to-ground recognition plans for any missions that require CAS or CCA support, especially if it involves nonstandard vehicles and HN personnel. The SFOD should coordinate with the aircrew to determine the method for visual identification and mark personnel and vehicles accordingly.

Favorable Weather

4-57. Unrestricted horizontal visibility improves CAS and CCA aircrew effectiveness regardless of aircraft type. Before CAS or CCA missions are executed, minimum weather conditions must be considered. The aviation component element commander determines the worst weather conditions in which CAS and CCA missions can be conducted based on regulations, aircraft and equipment limitations, and aircrew experience. Weather conditions worse than those considered to be the minimum will significantly degrade the ability to perform CAS and CCA because the majority of CAS and CCA aircraft do not have a true all-weather capability.

Prompt Response

4-58. To be truly effective, CAS and CCA must be responsive. Streamlined request and control procedures improve responsiveness. Prompt response allows a commander to exploit planned objectives and to take advantage of unanticipated battlefield opportunities. Techniques for improving response time include—

- Using the SOTF to decrease the distance to the AO.
- Placing aircraft on ground or airborne alert status.
- Delegating launch and diverting authority to subordinate units.

Joint Tactical Air Controller and Special Forces Operational Detachment Skill

4-59. CAS and CCA can range in complexity from very simple (such as controlling two aircraft) to complex (such as controlling several sorties of fixed- and rotary-wing aircraft and indirect fire). Aircrew and terminal controller skills have a direct influence on mission success. The SFOD commander, while planning CAS or CCA missions, must recognize the proficiency of his personnel. The SFOD should conduct CAS and CCA training as often as possible to maintain a high degree of skill.

Weaponeering

4-60. To achieve the desired level of destruction, neutralization, or suppression of enemy targets, the weapons load, arming settings, and fuse settings must be tailored for the desired results. For example, cluster and GP munitions are very effective against troops and stationary vehicles. Hardened, mobile, or pinpoint targets, however, may require specialized weapons, such as laser-guided, electro-optical (EO), or IR munitions, or aircraft with special equipment or capabilities. The requesting SFOD commander should solicit the type and quantity of ordnance that is appropriate to his needs and be informed of the ordnance load that each supporting CAS or CCA aircraft is carrying.

Communications and Information Systems

4-61. CAS or CCA execution requires dependable and interoperable communications. Unhindered voice or data communications between aircrews, air control agencies, JTACs, and the SFOD greatly increase the ease by which CAS or CCA is requested and controlled. Additionally, information flow will come from the battlefield in the form of in-flight reports and mission reports. Information systems that can relay timely and perishable information, such as target activity after attack and additional targets, will facilitate real-time CAS and CCA decision making, as well as future CAS and CCA planning.

Command and Control

4-62. CAS and CCA require integrated, flexible C2. C2 that facilitates an understanding of the mission and the initiative to adapt to changing battlefield situations is the foundation for creating conditions favorable for CAS and CCA employment. Basic requirements for CAS and CCA C2 are the ability to process requests, assign assets, communicate taskings, integrate fires and routings, coordinate support, establish airspace control measures, and update or warn CAS and CCA aircraft of enemy threats.

CONTROL AND COORDINATION PROCEDURES

4-63. The SFOD JTAC can assist the commander in requesting and employing a variety of measures to control and coordinate airspace in his AO through the airspace control authority. In joint operations, the airspace control authority receives and compiles requests for control and coordination locations and coordinates the airspace use by publishing the airspace control plan and the subsequent airspace control orders. The air and ground commanders coordinate the use of control procedures to strike a balance between the ground force use of airspace and protection of aircraft using that airspace. Control and coordination procedures include airspace coordinating measures (ACMs) and FSCMs.

4-64. ACMs increase operational effectiveness. They also increase CAS and CCA effectiveness by ensuring the safe, efficient, and flexible use of airspace. ACMs speed the handling of air traffic within the objective area, and terminal controllers use ACMs to control the movement of CAS and CCA aircraft over the battlefield. ACMs also include control points.

4-65. Control points route aircrews toward their target areas and provide a ready means of conducting fire support coordination (Figure 4-9, page 4-17). Control points should be easily identified from the air and should support the JFC's scheme of maneuver. Control points are the first variable the JTAC uses to manage the attack geometry for a given mission.

4-66. If possible, control points should be usable by a variety of aircraft. The JTAC identifies the specific use for each control point as the tactical situation dictates. The ATO SPINS state the intended use of control points. The following paragraphs identify multi-use control points:

- *Entry point and exit point.* Entry points and exit points are used to enter and exit the amphibious objective area (AOA). At entry points and exit points, the aircrew must contact the JTAC for further clearance.
- *En route point (ERP).* ERPs are used to define routes of flight to and from the target area. ERPs allow specific routing of aircraft for C2, airspace limitation, or ROE requirements. For the ingress routes, ERPs are placed between the rendezvous point and the contact point. For the egress routes, the ERPs are placed between the egress control point (ECP) and the penetration point (PP).
- *Orbit point and holding point (HP).* Orbit points and HPs either represent geographic positions or are defined and fixed by electronic means. These points are used to station aircraft inside the AO, keeping them in a specific area of airspace while they await further routing instructions. An orbit point is used during tactical operations when a predetermined pattern would be predictable and therefore is not set. An HP prescribes a predetermined pattern and is normally used while awaiting air traffic control (ATC) clearances. A control point can be dual-designated as both an orbit point and an HP, although orbit points and HPs are usually separate and distinct locations.
- *Contact point (CP).* JP 1-02 defines a CP as "the position at which a mission leader makes radio contact with an air control agency." During ingress, the aircrew contacts the JTAC at the CP. A CP allows coordination of final plans before entering heavily defended airspace.
- *Initial point.* JP 1-02 defines IP as "a well-defined point, easily distinguishable visually and/or electronically, used as a starting point for the bomb run to the target." IPs are located 5 to 15 nautical miles (nm) from the target area. JTACs and aircrews use IPs to help position aircraft delivering ordnance.

Tactical Facility Operations and Defense

- *Rendezvous point (RP).* An RP is a prearranged geographic location where aircraft meet after takeoff or after exiting the target area.
- *Egress control point (ECP).* An ECP is a well-defined geographical control point outside the enemy air defense area. The ECP identifies a CAS or CCA platform egress from the target. Contact with JTAC normally ends at the ECP.

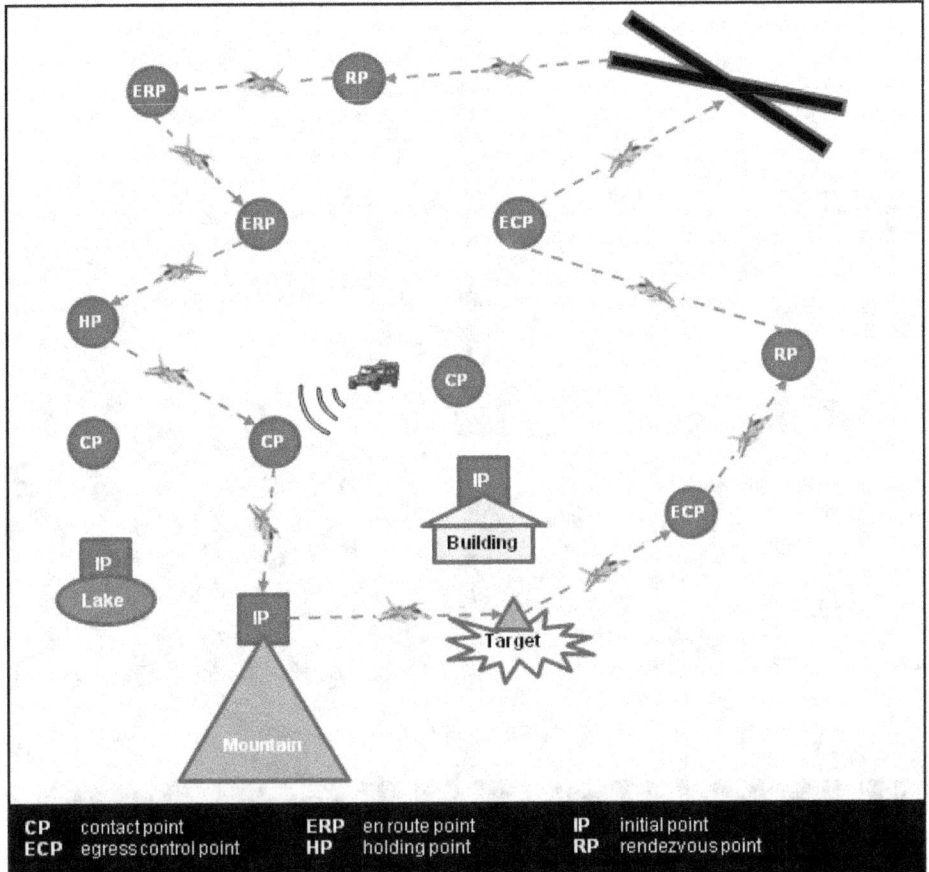

Figure 4-9. Fixed-wing airspace control measures

4-67. Although many multi-use control points apply to both fixed- and rotary-wing aircraft, there are two points that meet the unique requirements of rotary-wing attack platforms. The two points are holding areas (HAs) and BPs.

4-68. The HA is occupied while awaiting targets or missions. While in the HA, aircrews receive the CCA briefing and perform final coordination. Aircrews can receive updated target or mission information in a face-to-face brief or over the radio. After receiving the brief, aircrews move along attack routes to BPs or individual firing points. The HA should be well forward, yet provide cover and concealment from enemy observation and fires. The HA should be large enough for adequate dispersion and meet all landing zone (LZ) selection criteria. Because the HA is well forward, it may be necessary to supplement its security by

Chapter 4

providing it with a small security force. It must also be supplied with the necessary communications for coordinating a launch. The availability or use of messengers should also be considered and planned for.

4-69. BPs are maneuvering areas that contain firing points for attack helicopters. BPs should allow good cover and concealment, provide necessary maneuvering space, allow for appropriate weapons engagement zones, and be reasonably easy to identify.

FIRE SUPPORT COORDINATING MEASURES

4-70. The SFOD commander may employ permissive and restrictive FSCMs in his AO. The FSCMs are positioned and adjusted in consultation with superior, subordinate, supporting, and adjacent commanders. The SFOD commander requests FSCMs based on the mission requirements and the assets that are available to support the mission. FSCMs are used to facilitate timely and safe use of fire support and may be permissive or restrictive in nature. Figure 4-10 depicts common FSCMs.

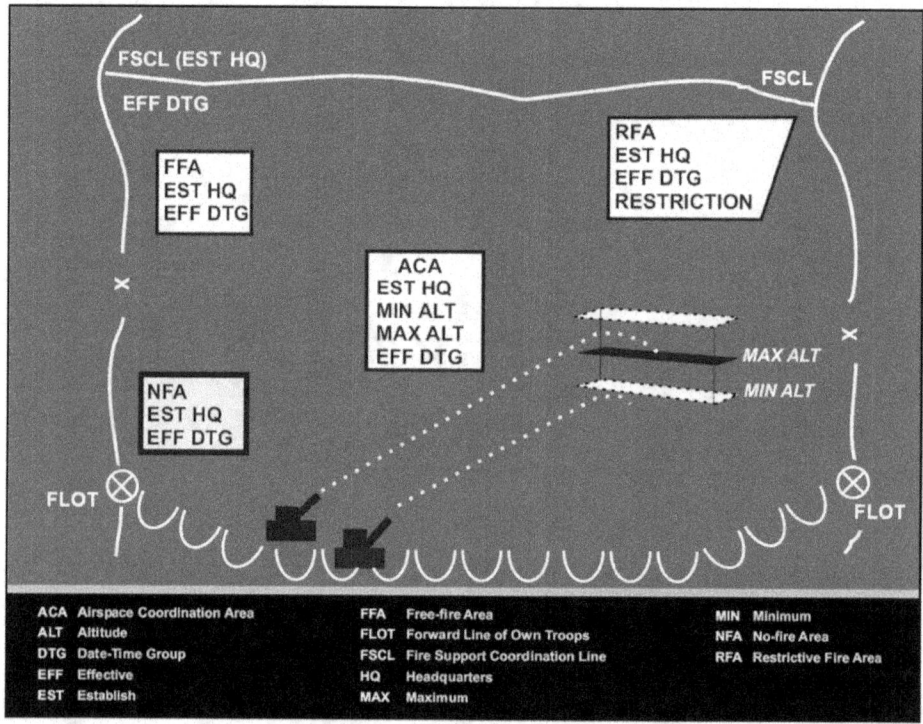

Figure 4-10. Formal airspace coordination area

Permissive Measures

4-71. Permissive measures facilitate target attacks. The three types of permissive FSCMs are as follows:
- *Coordinated fire line (CFL).* A CFL is a line beyond which conventional, indirect, surface fire support means may fire at any time within the boundaries of the establishing HQ without additional coordination.
- *Fire support coordination line.* An FSCL can be established and adjusted by the land or amphibious component commanders within their AOs in consultation with superior,

Tactical Facility Operations and Defense

subordinate, supporting, and affected commanders. FSCLs facilitate the expeditious attack of surface targets of opportunity beyond the coordinating measure. The FSCL applies to all fires of air-, land-, and sea-based weapon systems using any type of ammunition. Forces attacking targets beyond an FSCL must inform all affected commanders in sufficient time to allow necessary reaction to avoid fratricide. Supporting elements attacking targets beyond the FSCL must ensure that the attack will not produce adverse effects on, or to the rear of, the line. Short of an FSCL, the appropriate commander controls all air-to-ground and surface-to-surface attack operations. The FSCL should follow well-defined terrain features.

- *Free-fire area.* An FFA is a specific area into which any weapon system may fire without additional coordination with the establishing HQ.

Restrictive Measures

4-72. Restrictive measures safeguard friendly forces. There are numerous types of restrictive measures, to include the following:
- NFAs.
- RFAs.
- Restrictive fire lines (RFLs).
- ACAs.

No-Fire Area

4-73. An NFA is a land area designated by the appropriate commander into which fires and their effects are prohibited. There are two exceptions to this prohibition, described as follows:
- When establishing, HQ may approve fires temporarily within the NFA on a mission-by-mission basis.
- When an enemy force within the NFA engages a friendly force, the commander may engage the enemy to defend his force.

Restrictive Fire Area

4-74. An RFA is an area in which specific restrictions are imposed. Fires (or the effects of fires) that exceed those restrictions will not be delivered into the area without coordination with the establishing HQ.

Restrictive Fire Line

4-75. An RFL is a line established between converging friendly forces where one or both elements may be moving. The RFL prohibits fires or their effects across that line.

Airspace Coordination Area

4-76. An ACA is a three-dimensional block of airspace in a target area, established by the appropriate ground commander, in which friendly aircraft are reasonably safe from friendly surface fires. The two types of ACAs are—
- *Formal.* A formal ACA is established by the airspace control authority at the request of the appropriate ground commander. Formal ACAs require detailed planning. Although not always necessary, formal ACAs should be considered. The vertical and lateral limits of the ACA are designed to allow freedom of action for air and surface fire support for the greatest number of foreseeable targets. Because the supporting arms coordination center (SACC), fire support coordination center (FSCC), fire support element (FSE), or fire direction center (FDC) can determine the trajectory for a specific ground or naval surface fire support (NSFS) asset firing at a specific target, each target must be evaluated to ensure the trajectories of the rounds do not penetrate the ACA. The fire support coordinator (FSC) should consult the FDC when deciding the altitude of an ACA to determine if that altitude would allow the majority of targets to be attacked without interference or problems. Formal ACAs are promulgated in the airspace control order (ACO) or the ATO SPINS.

Chapter 4

- *Informal.* An informal ACA can be established using separation plans and may be established by any ground commander. Aircraft and surface fires may be separated by—
  - *Lateral separation.* Lateral separation is effective for coordinating fires against targets that are adequately separated from flight routes to ensure aircraft protection from the effects of friendly fires.
  - *Altitude separation.* Altitude separation is effective for coordinating fires when aircraft remain above or below indirect fire trajectories and their effects.
  - *Altitude and lateral separation.* Altitude and lateral separation is the most restrictive technique for aircrews and may be required when aircraft must cross the firing unit's gun-target line.
  - *Time separation.* Time separation requires the most detailed coordination and may be required when altitude restrictions from indirect fire trajectories adversely impact aircraft ordnance delivery (for example, mortar trajectory).

Chapter 5
Tactical Facility Sustainment

When the theater support system is in place, it can meet most SF unit TACFAC requirements. TACFAC planners then must concentrate on the type of sustainment required, the number of days of accompanying supplies based on the time-phased force and deployment list (TPFDL), and the SF unit TACFAC basing needs.

Each TACFAC sustainment operation is unique and requires mission-specific analysis that develops a tailored sustainment force. Joint, international, interagency, and NGO activities add complexity to the TACFAC sustainment system. Because of their geographic location, SF units may conduct operations outside a theater support system. Preparing and submitting a detailed SOR during these types of missions can enhance the unit's priority determination process and also add a final coordination check to the theater operation plan (OPLAN).

LOGISTICS OVERVIEW

5-1. The SOTF SPTCENs provide or coordinate through the joint special operations task force (JSOTF) for SF TACFAC logistics on a unit-sustainment basis for all elements assigned or attached to their respective facilities. SF logistics planners and personnel apply their knowledge of conventional logistics operations to meet the specific TACFAC requirements generated by SF units. Logistics fundamentals apply to most SF operations. TACFAC logistics support normally includes—

- Requisition, receipt, storage, and distribution of all classes of supply.
- Procurement of nonstandard supplies and items of materiel for TACFAC improvement and sustainment.
- TACFAC protection and security items, to include obstacles, barriers, and wire.
- Mortuary affairs.
- Production and distribution of food and potable water.
- Field service, supply, and repair (SSR) maintenance for all wheeled vehicles, power-generation equipment, signal equipment, diving and marine equipment, and small arms.
- Limited SSR for SF-peculiar equipment.
- Airdrop equipment rigging, supply, and repair.
- Force health protection (FHP).
- Human resources (HR) support.

DEVELOPED THEATER LOGISTICS

5-2. In a developed theater, a sustainment base sets up within the theater. Pre-positioned war reserve materiel stocks (PWRMSs) and operational project stocks are in place and foreign nation support (FNS) agreements exist. The SPTCEN in a developed theater supports the following logistics TACFAC functions:

- Supply.
- Field services.
- Maintenance.
- Transportation.

Chapter 5

SUPPLY

5-3. The service detachment's supply and transportation section requisitions, receives, and stores standard Class I, II, III, IV, VI, and VII supplies from the supporting supply and service company. All of these classes of supplies (except bulk Class III) are demand items. The using TACFAC submits a request through the service detachment. Field maintenance either fills the request from its existing stocks or forwards the request to the supporting unit. Field maintenance uses a combination of supply point, unit, and throughput section requisitions and receives Class VI packages the same way they requisition Class I supplies.

5-4. Bulk Class III is a scheduled item. The JSOTF logistics officer forecasts unit and TACFAC requirements through logistics channels to the theater sustainment command (TSC) or sustainment brigade based on input from the battalions. The TSC deputy chief of staff for logistics (DCSLOG) and theater Army material management command (TAMMC) develop a distribution plan to allocate fuel to subordinate units and TACFACs based on fuel availability (IAW theater OPLANs) and unit priorities.

5-5. The supply and transportation section requisitions and receives nonstandard, SF-peculiar items (such as large-capacity diesel generators) through the Army special operations forces liaison element (ALE). The ALE fills the request from the theater or—in the case of certain non-DOD items—obtains the items through the special operations command (SOC) logistics staff section (J-4).

5-6. The supply and transportation section requests, draws, and stores conventional Class V supplies from the supporting ammunition supply point (ASP). A conventional ordnance ammunition company of the TSC ammunition group operates the ASP and uses supply point distribution. Class V supply is scheduled, not demanded. Based on input from the battalions, the SFG(A) operations staff section (S-3) must determine the group's operational requirements, unit basic load (UBL), and required supply rate. The S-3 then submits the requirements through operational channels for approval and allocation by the TSC deputy chief of staff for operations (DCSOPS). DCSLOG and TAMMC allocate scarce Class V items by computing a controlled supply rate based on guidance from the Army Service component command (ASCC) DCSOPS. Once the SFG(A) commander receives the Class V allocation, the SFG(A) and battalion S-3s approve unit Class V requests before the logistics staff sections (S-4s) can fill them.

5-7. The group or battalion medical section requisitions and receives its normal Class VIII supplies from the supporting medical treatment facility of the medical deployment support command (MDSC). The medical facility uses a combination of unit and supply point distribution. Class VIII resupply is on demand. The using TACFAC submits requests to the medical supply sergeant, who forwards the request through medical channels to the medical facility. The facility either fills the request from its existing stocks or forwards the request to its supporting medical logistics (MEDLOG) unit. For bulk issue of Class VIII supplies to fill SF operational requirements, the MDSC normally authorizes direct requisitioning from the MEDLOG unit.

5-8. The service detachment's mechanical maintenance section requisitions, receives, and stores Class IX supplies from supporting field maintenance in the sustainment brigade. Class IX resupply is on demand. The using TACFAC submits its request to the mechanical maintenance section. It forwards the request to field maintenance. Field maintenance fills the request from its existing stocks or forwards the request to the TAMMC.

5-9. The supply and transportation section receives and stores Class X supplies from the supporting TSC. The TSC uses a combination of unit, supply point, and throughput distribution. The using TACFAC submits its request through the base plans staff section (S-5). The S-5 forwards the request through logistics channels.

5-10. The battalion supply and transportation section obtains potable and nonpotable water from local sources using organic equipment. When water requirements exceed the local supply, the section requisitions and draws water from a water supply point set up by the supporting field maintenance supply and service company.

5-11. The battalion supply and transportation section requisitions and receives unclassified maps from the supporting field maintenance unit. Field maintenance obtains unclassified maps from the appropriate ASCC map depot. Using units and TACFACs submit their requests to the intelligence staff section (S-2),

Tactical Facility Sustainment

who then consolidates them and forwards the requests through supply channels. The battalion S-2 requisitions and receives classified maps and other classified intelligence products through intelligence channels.

5-12. To meet their operational requirements during the transition to active operations and during unanticipated breaks in normal resupply operations, TACFACs maintain UBLs of Class I, II, III, IV, V, VII, and IX supply items. TACFAC commanders should review these UBLs at least annually to ensure that they adequately address current operational sustainment requirements. TACFAC commanders inspect their UBLs periodically for proper maintenance, rotation, and security. Commanders also ensure adequate spare parts and necessary equipment has been requisitioned.

FIELD SERVICES

5-13. Field services include mortuary affairs, airdrop (Figure 5-1), clothing exchange and bath, laundry, bread baking, textile and clothing renovation, and salvage. Mortuary affairs and airdrop are primary field services because they are essential to the sustainment of combat operations. All others are secondary field services.

Figure 5-1. Airdrop resupply

5-14. Whenever possible, TACFACs that sustain fatal casualties identify the human remains and place them in human-remains pouches. The SFOD then evacuates the remains to the service detachment for further evacuation to the supporting mortuary affairs collection point. If the remains are contaminated, the remains and the pouches should be so marked. When an SFOD cannot evacuate its dead, it conducts an emergency burial and reports the burial to the battalion. The battalion S-4 submits a record of interment through mortuary affairs channels. Whenever possible, a unit chaplain or the TACFAC commander conducts an appropriate service to honor the dead.

Chapter 5

5-15. The SOTFs may not have fixed facilities or civilian contractors to provide secondary field services. In this situation, the field maintenance supply and service company provides these services when the situation permits.

MAINTENANCE

5-16. The service detachment's mechanical maintenance section performs limited consolidated unit-level maintenance of wheeled vehicles and power-generation equipment. It performs vehicle recovery (Figure 5-2). The signal detachment's electronic maintenance section performs field maintenance of signal equipment. It also performs limited sustainment maintenance of SF-peculiar signal equipment. Unit armorers perform field maintenance of small arms.

Figure 5-2. Vehicle recovery

5-17. Required maintenance on an item of equipment may exceed TACFAC capabilities. In such instances, the mechanical maintenance section, electronics maintenance section, or TACFAC evacuates the equipment to the supporting field maintenance company or requests on-site repair by a mobile maintenance support team from that company.

5-18. There are exceptions to these procedures. The rigger air delivery section, for example, evacuates unserviceable airdrop equipment to the TSC airdrop equipment repair and supply company. Similarly, the medical section evacuates unserviceable medical equipment to the supporting medical treatment facility or MEDLOG unit.

5-19. For those items of SF-peculiar equipment the Army maintenance system cannot repair, the SFG(A) must rely on the group support battalion (GSB) or on civilian specialists and TACFAC personnel who have attended civilian maintenance training. Such equipment may require evacuation to the continental United States (CONUS) for repair at the manufacturer or other selected facility.

Tactical Facility Sustainment

TRANSPORTATION

5-20. The primary concern of the service detachment commander is transportation operations (air, motor, rail, and water). The supply and transportation section provides the trucks to support supply point distribution and other normal logistics support activities. It does not, however, have dedicated drivers for these trucks. The TACFAC commander may organize a provisional transportation section by assigning drivers to these trucks. The TSC transportation section may attach appropriate motor and water transportation assets to the support company for atypical logistics support operations. Otherwise, transportation units support unusual transportation requirements on a mission-priority basis with their sustainment assets. The unit S-4 coordinates for transportation support through the regional transportation management office (TMO) of the theater Army movement control agency (TAMCA). When the same TMO services the SOTF, the TMO may require the group S-4 to consolidate support requests.

UNDEVELOPED THEATER LOGISTICS

5-21. An undeveloped theater does not have a significant U.S. theater sustainment base. PWRMSs, in-theater operational projects, and FNS agreements are minimal or nonexistent. When an SFOD deploys to a TACFAC in an undeveloped theater, the unit must bring sufficient resources to survive and operate until the ASCC can establish a bare-base support system or make arrangements for HN and third-country support. The bare-base support system may function from CONUS, afloat (amphibious shipping or mobile sea bases), or at a third-country support base. The bare-base support system relies heavily on strategic airlift and sealift for resupply.

LOGISTICS SUPPORT OPERATIONS

5-22. Deployed SF units in an undeveloped theater TACFAC occasionally bypass normal logistics support echelons. They may maintain direct contact with their parent units in CONUS, or they request a tailored support package from the GSB to accompany them into the theater. The GSB can then request directly from the CONUS wholesale logistics system (through the GSB) and provide support and sustainment to the SF units. They also rely on ASCC contracting and CA expertise to obtain support and sustainment. Usually, the solution is some combination of all four options.

SUPPORT RELATIONSHIPS

5-23. Support relationships must be developed and nurtured before and during exercises, mobile training teams (MTTs), joint combined exchange training (JCET), and planning conferences. Support relationships identified in the theater support plan are a basis for continued support relationships between the JSOTF and the ASCC elements providing its support package. The support package should be provisionally organized as a composite support battalion or company.

SUPPLY

5-24. Normal basic loads are inadequate for TACFAC operations in an undeveloped theater. SF units deploying into an undeveloped theater should recalculate requirements based on the TACFAC mission. For example, an SFODA may deploy with 30 days of supply (15-day order-ship time, 10-day operating level, 5-day safety level). Because this quantity exceeds the SFG(A) capacity to move and store supplies, the group and battalion S-4s normally divide the loads into accompanying supplies and preplanned follow-on supplies. Accompanying supplies normally are limited to the unit's basic and prescribed loads, plus additional Class I, III, and V supplies critical to the operation. The group and battalion S-3s must include accompanying supplies in all predeployment planning.

5-25. Supply procedures vary in an undeveloped theater. The JSOTF can rely on local contract support for fresh Class I supplies and DFAC operation. The SFODA in a TACFAC routinely purchases Class II, III, IV, and VI supplies locally or from third-party contractors. The SFODA normally receives Class V and IX supplies through the standard U.S. system. The TACFAC unit stocks low-density, high-dollar repair parts not normally authorized at unit-maintenance level. Class VII supplies may include a combination of military and commercial equipment from U.S. and foreign sources, particularly large-capacity diesel

Chapter 5

generators. Replacement of equipment depends on the duration of the operations, theater sustainment and maintenance repair capabilities, loss rates, and the availability of operational readiness float or PWRMS. The SOTF and TACFAC contracts or procures water locally.

FIELD SERVICES

5-26. The JSOTF normally receives field services through the GSB until the TSC establishes these capabilities. The SFG(A) may contract for various housekeeping services, including laundry services. If laundry services are unavailable, the group S-4 will arrange for clothing exchange through the GSB.

MAINTENANCE

5-27. Preventative maintenance checks and services (PMCS) are critical in tropical, desert, or arctic environments that typically exist in undeveloped theaters. The frequency of periodic services often increases in these regions. Repair facilities in an undeveloped theater are often unreliable and unavailable. The JSOTF commander should review the modified table of organization and equipment (MTOE) to determine the items he needs to meet increased maintenance demands caused by operations in an undeveloped theater. For example, the commander may need repair parts, special tools, or diagnostic equipment for testing commercial off-the-shelf (COTS) diesel generators. The TACFAC unit should identify maintenance support in their SOR before deployment. The TACFAC commander also may contract for HN maintenance support of its equipment.

TRANSPORTATION

5-28. Because undeveloped theaters have poor LOCs, Army aviation assets deploy early (whenever possible) to support SF logistics operations. These aviation assets must include an adequate maintenance support package for autonomous, unilateral, and continuous operations. The battalion commander reviews HN (or any other) lift assets to meet additional unresourced transportation requirements. Regardless of the source of aviation assets used to support a JSOTF, this support must be dedicated for administrative and logistical sustainment requirements in both mature and undeveloped theaters.

HUMAN RESOURCE SUPPORT

5-29. HR support remains essentially unchanged in an undeveloped theater. The ASCC, in coordination with the SOC, develops personnel replacement plans.

FORCE HEALTH PROTECTION

5-30. The SFG(A) can deploy with an FHP package to provide dedicated support until normal TSC health services are set up. The SFG(A) has extensive organic medical capabilities. At the group level, a flight surgeon, dental officer, veterinary officer, medical operations officer, medical logistics officer, and environmental officer are all assigned to the SOTF and are available to the SF TACFAC medic. At the battalion level, each SOTF has authorization for a flight surgeon and a physician assistant. At the SOTF, the surgeon and physician assistant can perform advanced trauma life-support procedures and provide limited resuscitative care. The FHP package also includes a preventative medicine NCO capable of providing medical threat evaluation and limited direct preventative medicine support. The lowest level of the FHP package is the 18D as the SF TACFAC health-care provider.

5-31. In an undeveloped theater, the JSOTF surgeon may use U.S., HN, or third-country medical facilities during normal operations to augment the medical capabilities of the group and battalion medical sections. In this case, a group or battalion aid station may be set up away from the SOTF in a centrally located HN hospital or clinic supporting multiple deployed SFODAs and their TACFACs. Vehicular casualty evacuation (CASEVAC) to the SOTF is unlikely because of the considerable distances that normally separate the SFODAs from the larger bases or other U.S. medical support. Helicopter CASEVAC is the preferred method.

SPECIAL FORCES OPERATIONAL DETACHMENT SUPPORT AND SUSTAINMENT

5-32. All SFODAs and TACFACs require services to sustain food, water, and clothing, as well as medical and personnel needs. SFODAs may use a combination of TSC logistics support, organic support companies, or other logistics support systems to maintain their operations. SF commanders and their staffs task-organize their assets to work with and employ the logistics support procedures and mechanisms existing in the theater.

5-33. TACFAC sustainment operations conducted with deliberate planning adhere to normal logistics support operations. Mission planners must consider the theater—
- Medical capabilities.
- Transportation and petroleum, oils, and lubricants (POL) capabilities.
- Resupply capabilities.
- Repair capabilities.

5-34. SF TACFAC commanders brief SOTF commanders and staffs on the quantity and types of equipment and supplies that will accompany the SFODA during infiltration. Factors that influence the selection of the accompanying supplies include the following:
- METT-TC and terrain analysis.
- Size and capability of the HN force and its logistics needs.
- Availability of resources in the AO.
- Method of infiltration and exfiltration.
- Operational posture.
- Degree of difficulty to repair or replace critical MTOE and TACFAC items in the AO.

5-35. Based on these considerations, the SOTF staff sets up TACFAC supply levels for each class of supply in the joint special operations area (JSOA). It then determines the sequence, method, and timing of delivery. The SOTF plans for the following four types of TACFAC resupply operations:
- *Automatic resupply.* Automatic resupply provides items that do not accompany the SFODA during infiltration. Automatic resupply provides sustainment, training, and operational supplies to the SFODA and indigenous forces on a preset schedule. The delivery time, location, contents, identification marking system, and authentication are preplanned. The SOTF sends supplies automatically unless the SFODA cancels, modifies, or reschedules the delivery.
- *Emergency resupply.* Emergency resupply provides mission-essential equipment and supplies to restore operational capability and survivability to the SFODA and its indigenous force. The SOTF delivers an emergency resupply when one of the following occurs:
  - Radio contact has not been established between the deployed SFODA and the supporting SOTF within a set time after infiltration.
  - The deployed SFODA fails to make a preset consecutive number of scheduled radio contacts to the SOTF.
- *On-call resupply.* On-call resupply provides equipment and supplies to a deployed SFODA and its TACFAC to meet operational requirements that cannot be carried during infiltration, or to replace equipment lost or damaged during the operations. The deploying unit, rigger section, and S-4 prepack on-call resupply bundles. The bundles are held in a secure location and then delivered upon SFODA request. SF use the catalog supply system (a brevity code system) to expedite on-call resupply requests, to ensure accurate identification of supply items, and to minimize message length. The catalog supply system lists equipment and supplies by class of supply. It groups associated equipment and supplies into conventional unit sets. It then assigns code words to each catalog item and set. The SOC J-4 prepares the theater supply catalog, and the SOC command, control, communications, and computer systems staff section (J-6) reproduces it as a signal operating instruction (SOI) item.

- *Caches.* A cache is an alternative form of resupply. SFODAs can stockpile materiel within the JSOA to support future operations. They also can use caches emplaced by other units on previous operations. Using other unit's caches from previous operations must be coordinated and approved by the JSOA commander.

5-36. The SOTF S-4 requests the supplies and equipment for the SFODA missions through the GSB to the TSC. Resupply missions are normally preplanned by the SFODAs while in isolation. An SFODA at an isolated TACFAC may be required to obtain supplies from a conventional force due to a challenging geographic location. SF leaders should strive to establish a good working relationship with nearby conventional units.

5-37. The GCC establishes the command relationship involving ARSOF in the theater. However, the theater ASCC has Title 10, United States Code (10 USC) responsibility—regardless of operational control (OPCON) arrangements within the combatant command—to provide administration and support to deployed ARSOF. Also, when directed by the GCC, the ASCC will support and sustain designated SOF of other U.S. Services and other multinational SOF.

5-38. Special operations support command (SOSCOM) HQ provides C2 to its organic elements to accomplish planning and coordinate health and signal support to ARSOF units supporting the combatant commanders (CCDRs). When directed, SOSCOM deploys its battalions in direct support of deployed ARSOF, such as SF TACFACs.

5-39. The special operations theater support element (SOTSE) is a staff planning, coordinating, and facilitating element. It serves as the ARSOF liaison to the ASCC for matters pertaining to logistics and medical needs, and provides ARSOF advocacy within the ASCC. The SOTSE coordinates requirements identified by ARSOF and facilitates the interface of ARSOF organizational logistics functions with the services provided by the ASCC. USASOC attaches the SOTSE to the ASCC HQ for duty within the ASCC logistic staff. The SOTSE coordinates closely with the supported theater special operations command (TSOC) and ARSOF during the deliberate planning process. The SOTSE identifies support requirements, integrates ARSOF sustainment requirements into the ASCC support plan, and ensures timely provision of that support.

5-40. A critical source of information that the ASCC needs for coordination and facilitation functions is the SOR provided by the SF units. The SOC J-4 and other logistics staffs have to be proactive and must be included in the mission-planning process. The logistics planners must anticipate operational unit requirements at all stages of the mission. Ideally, the J-4 uses the ASCC OPLAN in preparing the concept plan (CONPLAN) for inclusion in the mission order. This approach allows theater support elements time to review required support before the SOF mission unit submits its mission-tailored SOR. This review is especially critical in crisis action planning (CAP) and short-notice mission changes. The SOR (Figure 5-3, pages 5-9 through 5-15) is a living document that requires periodic reevaluation and updating as the TACFAC requirements change. Determination of requirements begins with the receipt of the mission.

5-41. There are 10 established numbered classes of supply and one miscellaneous category of supply. All classes of supply are directly related to support and supply the TACFAC construction and sustainment. Only two categories of supply—Classes VI and VIII—may not directly apply. The classes of supply (with examples) are as follows:
- *Class I.* Class I supplies are subsistence items and gratuitous health and welfare articles, such as meals, ready to eat (MREs) and fresh fruit and vegetables.
- *Class II.* Class II supplies include equipment authorized and allowed to the unit, such as clothing, personal tools, and administrative supplies.
- *Class III.* Class III supplies are either bulk or packaged POL. Bulk POL include gasoline and multifuels. Packaged POL include coolant, de-icer, and antifreeze compounds.
- *Class IV.* Class IV supplies are construction and barrier materials, to include all fortifications, lumber, sandbags, and barbed wire.
- *Class V.* Class V supplies are ammunition of all types, explosives, mines, fuzes, detonators, pyrotechnics, and chemical and nuclear munitions.

- *Class VI.* Class VI supplies are personal demand items, usually purchased at the post exchange (PX), to include soap, toothpaste, alcohol, and cigarettes. This supply class is usually requisitioned and distributed with Class I items.
- *Class VII.* Class VII supplies are major end items, including aircraft, tanks, major weapon systems, and vehicles.
- *Class VIII.* Class VIII supplies are medical items, to include medicine, equipment, and repair parts peculiar to medical equipment. The two subclasses of this Class VIII supplies are—
  - *Class VIIIa*, which includes medical consumable supplies, not including blood or blood products.
  - *Class VIIIb*, which includes blood and blood products, to include whole blood and blood plasma.
- *Class IX.* Class IX supplies are repair parts and components, to include assemblies and subassemblies (whether they are repairable or not), required for maintenance support of all equipment, such as spark plugs and batteries.
- *Class X.* Class X supplies are materials required to support nonmilitary programs which are not included in Classes I through IX, such as agriculture and economic development.
- *Miscellaneous.* The miscellaneous category of supplies includes water, salvage, and captured materiel.

(CLASSIFICATION)

I. REFERENCES.

II. GENERAL.
   A. Unit to be supported.
   B. When support is required.
   C. Location of supported unit when support is required.
   D. Unit points of contact (POCs).
   E. Number of personnel to be supported.
   F. Unit identification (ID) code.
   G. Force activity designator.
   H. Funding. Special funding for the operation and how to access, if applicable, and fund flow for obtaining supplies, including project code.

III. CONCEPT OF OPERATIONS.
   A. Mission. State the general mission of the unit, command, or operation.
   B. Desired results. Provide a concise statement of the desired results of the support being requested.

IV. ASSUMPTIONS. Give the conditions that are likely or must exist for this support to be required. Relate the assumptions to specific requirements, as required or appropriate.

V. CONSTRAINTS. Define situation that, if experienced, will degrade operations. Give conditions to specific requirements identified, as required or appropriate.

VI. COMMAND, CONTROL, AND COORDINATION. Describe functional C2 of the unit.

VII. SUPPLIES.
   A. *Class I.*
      1. DFAC requirements.

(CLASSIFICATION)

Figure 5-3. Statement of requirements format

Chapter 5

(CLASSIFICATION)

    2. Augmentation.

    3. Food storage facilities. Determine which of the following food storage facilities are required to contain a 30-day supply of rations.

        a. Dry space (in cubic feet).

        b. Chill space (in cubic feet).

        c. Freezer space (in cubic feet).

    4. Mermites. Determine requirements for mermites. List how many and how often they are required.

    5. Meal payment. Determine how individuals will pay for their meals.

        a. Cash collection.

        b. Payroll deduction.

        c. Meal cards.

    6. DFAC hours. Determine if a 24-hour DFAC will be required.

    7. Equipment augmentation. Determine requirements for equipment augmentation to dining facility. List the equipment by nomenclature, National Stock Number (NSN), and quantity.

    8. Combat rations. Estimate the number of combat rations required by the number of meals required for 30-day sustainment.

        a. Meal, combat, individual.

        b. MRE.

        c. Long-range reconnaissance patrol (LRRP) rations.

        d. Other (specify).

    9. Water requirements.

    10. Other (specify).

  B. *Class II.*

    1. Self-service. List essential self-service supply center items required for 30-day sustainment.

    2. CBRN equipment. List requirements for CBRN consumables and nonconsumables needed to provide two complete issues of CBRN equipment following a CBRN attack.

    3. Sustainment. List other Class II items required for sustainment, such as items listed in Common Table of Allowance (CTA) 50-900, *Clothing and Individual Equipment.*

    4. Reproduction equipment. Determine what reproduction equipment is required. List the equipment and the number of copies needed for 30-day sustainment.

    5. Special equipment. List any special Class II equipment required over and above that already authorized and on hand. List by nomenclature, NSN, and quantity.

    6. Clothing sales. Determine requirements for clothing sales facility.

    7. Geospatial products (maps, digital data). State the requirement for each product type by quantity and area coverage.

    8. Other (specify).

  C. *Class III.*

    1. POL. Determine POL, including base-support functions, for a 30-day sustainment. List item by type and quantity.

        a. Motor gasoline (specify regular or super).

        b. Diesel fuel (specify DF1 or DF2).

(CLASSIFICATION)

Figure 5-3. Statement of requirements format (continued)

Tactical Facility Sustainment

(CLASSIFICATION)
- c. Aviation gasoline (specify JP4 or JP5).
- d. Oil (bulk).
- e. Grease.
- f. Coolants.
- g. Packaged POL or other lubricants.
2. Tankers and dispensers. Identify requirements for tankers or dispensers in addition to organic capabilities. List by type, capacity, and quantity.
3. Planning factors. Determine if the planning factors used to identify POL requirements were factors other than those in the United States Combined Arms Support Command (CASCOM) database or operational logistics (OPLOG) planner. If so, specify.
4. Other (specify).

D. *Class IV.* Determine requirements for building and barrier materials.
1. Determine requirements for building and barrier materials.
2. Other (specify).

E. *Class V.*
1. Additional Class V requirements. Determine Class V requirements over and above those in the UBL. List by DOD identification code, nomenclature, and quantity.
2. Planning factors. Determine which planning factor was used to forecast Class V consumption rates.
3. Other (specify).

F. *Class VI.*

G. *Class VII.*
1. Additional equipment. Determine requirements for additional items of equipment, such as trucks and generators. List by nomenclature, NSN, and quantity.
2. Maintenance augmentation. Determine the requirement for maintenance augmentation to support the equipment listed above.
3. Other (specify).

H. *Class VIII.*
1. Determine requirements for Class VIII supplies by nomenclature, NSN, quantities, and special requirements associated with a particular item, such as refrigeration.
2. Determine schedule of resupply required.
3. Determine whether resupply will be prepackaged standard line items.
4. Identify Class VIII supplies peculiar to the AO and whether they are readily available or must be specifically acquired (for example, antivenins).
5. Determine availability and reliability of HN Class VIII for emergency purposes.
6. Other (specify).

I. *Class IX.*
1. Mandatory parts list. Determine if there is a mandatory parts list to support the equipment.
2. Prescribed load list (PLL). Determine if the PLL includes repair parts to support all assigned equipment.
3. Equipment density list. Develop an equipment density list to provide to HN or other supporting agency, as required.

(CLASSIFICATION)

Figure 5-3. Statement of requirements format (continued)

(CLASSIFICATION)

    4. Leased vehicles and equipment. Determine Class IX requirements for leased vehicles and equipment, if necessary.

    5. Other (specify).

  J. *Class X.*

    1. Determine Class X requirements.

    2. Type.

    3. Quantity.

    4. Other (specify).

  K. *Other* (water, salvage, and captured materiel).

    1. Emergency resupply. Identify requirements for emergency resupply push packages. (Specify by NSN, nomenclature, and quantity. Attach as separate enclosure for each type of push package.)

    2. Maps and photographs. Identify requirements for maps and imagery.

    3. Other (specify).

VIII. **SERVICES.**

  A. Field services.

  B. Engineering services.

    1. Equipment power compatibility. Determine the following if supplied with commercial power at the wartime site.

      a. Equipment is compatible.

      b. Plug adapters are required. List what voltage and how many are needed.

      c. Transformers are required. List what voltage and how many are needed.

    2. Water requirements. Identify daily requirements for potable water.

    3. Pest control requirements. Determine requirements for rodent and insect control assistance.

    4. Heavy-equipment requirements.

    5. Other (specify).

  C. Other services.

    1. Linen requirements. List by type and quantity.

    2. Laundry services requirements.

    3. Other services identification. Determine if other services are needed.

IX. **MAINTENANCE.**

  A. Direct support (DS) and general support (GS) maintenance. Identify requirements for DS and GS maintenance.

  B. Other maintenance equipment. List commercial and nonstandard equipment that must be maintained.

X. **TRANSPORTATION.**

  A. Air transportation.

    1. Unit load plans. Enclose unit load plans.

    2. Administrative aircraft. Determine requirements. Specify type and number of hours per week.

(CLASSIFICATION)

Figure 5-3. Statement of requirements format (continued)

(CLASSIFICATION)

    3. Equipment and personnel requirements. Determine requirements for additional materiel handling equipment and personnel.

    4. Cargo storage facilities. Determine requirements for cargo storage facilities. Specify by the number of square feet required for the following:

        a. Covered secure storage.

        b. Outdoor secure storage.

    5. Airfield requirements.

        a. C-130s.

        b. Other (specify).

  B. Water transportation. Determine water transportation needs (specify).

  C. Ground transportation. Determine requirements for supplemental military vehicles.

    1. Tactical vehicles (with or without communications equipment).

    2. Other special-purpose vehicles.

XI. FACILITIES.

  A. Maintenance facilities. Identify vehicle, communication, weapons, and aviation maintenance area (covered) requirements (list in square feet).

  B. Billeting facilities.

    1. Billet number and size requirements.

        a. Officers.

        b. Senior enlisted.

        c. Enlisted.

        d. Females.

    2. Tentage.

  C. Medical facilities. Determine requirements for physical facilities.

    1. Hospital beds.

    2. Treatment rooms.

    3. Dental treatment rooms.

    4. Laboratory.

    5. X-ray room.

    6. Pharmacy.

    7. Other (specify).

  D. Other facilities (list by function and square feet).

    1. OPCEN.

    2. SPTCEN.

    3. SIGCEN.

    4. ISOFAC.

    5. DFAC.

    6. Dispensary.

    7. Reception and palletizing.

    8. Parachute rigging and drying.

    9. Administration.

(CLASSIFICATION)

Figure 5-3. Statement of requirements format (continued)

Chapter 5

> **(CLASSIFICATION)**
>
>   10. Unexploded ordnance (UXO).
>   11. Ammunition storage.
>   12. Fuel storage and refueling.
>   13. MWR (television room, telephone, Internet, weight room).
>   14. HN meeting room and classroom.
>   15. Gym.
>   16. Antenna fields.
>   17. Ranges (list type weapons requiring ranges).
>   18. Drop zones (DZs).
>   19. Secure facilities (for storing, receiving, and transmitting classified messages and documents).
>   20. Other (specify).
>
> **XII. PERSONNEL SERVICES.**
>   A. Casualty reporting. Determine how casualty-reporting system works.
>   B. Administrative services.
>   1. Reproduction and word processing. Determine reproduction and word-processing requirements.
>   2. Postal. Identify postal requirements.
>   C. Finance. Determine finance support requirements. Identify what is required (pay and allotments).
>   D. Religious support. Identify religious support requirements.
>   E. Legal. Determine requirements for staff judge advocate (SJA) support.
>   F. Other (specify).
>
> **XIII. MEDICAL.**
>   A. Patient care. Determine legal and policy constraints on providing medical care to indigenous personnel.
>   B. CASEVAC. Determine aeromedical and overland evacuation requirements.
>   C. MEDLOG.
>   D. Medical intelligence (MEDINT).
>   E. Preventive medicine services. Vaccination and prophylactic medication requirements.
>   F. Veterinary services.
>   G. Dental services.
>   H. Laboratory services.
>   I. Other (specify).
>
> **XIV. SIGNAL.**
>   A. Terminal equipment and access. Determine requirements for the following:
>   1. Supplemental terminal equipment. Specify by type and quantity.
>   2. Access to HN communication telephone system. Specify need, such as number of lines.
>   3. Access to North Atlantic Treaty Organization (NATO) telegraph network.
>   4. Access to HN military teletype system.
>   5. Access to automatic secure voice communications.
>
> **(CLASSIFICATION)**

Figure 5-3. Statement of requirements format (continued)

(CLASSIFICATION)

    6. Access to NATO secure voice network.
    7. Access to Automatic Digital Network (AUTODIN).
    8. Other (specify).
- B. Transmit and receive sites. Determine the number of transmit and receive sites to be set up and how much area will be needed.
- C. Signal maintenance support.
- D. Frequency requirements.
- E. Other (specify).

XV. SECURITY.
- A. MP. Determine MP requirements.
- B. CI. Determine if CI is required.
- C. Base defense. Determine if base-defense capabilities are required.
- D. Other (specify).

(CLASSIFICATION)

Figure 5-3. Statement of requirements format (continued)

This page intentionally left blank.

Chapter 6

Tactical Facility Funding

The SF field ordering officer (FOO) and paying agent (PA) officer must address competing needs, requirements (such as weapons systems, C2 systems, security, and logistics), and their demands on resources. SF unit resources include time, funds, personnel, existing contractor support, existing equipment and material, and other items and assets used to accomplish the mission. Any shortage of resources normally results in contracting having to meet the mission and fulfill necessary requirements. Each requirement has an associated cost—either in money or resources—which must be included in the budget or operations plan. When accurate preplanning is conducted, the FOO and PA have a budget to work with and can purchase necessary supplies or services through contracting. The needs dictated by a requirement are theater-unique and change often.

FUNDING PRINCIPLES

6-1. In the United States Central Command (USCENTCOM) area of responsibility (AOR), the Army is in charge of contracting through the Assistant Secretary of the Army for Acquisition, Logistics, and Technology (ASAALT). The steps in the acquisition process are as follows:
- The unit-generated requirement is documented using the following:
  - Statement of work (SOW) and an independent government cost estimate (IGCE).
  - DA Form 3953 (Purchase Request and Commitment [PR&C]).
- The requirements package is validated by the joint acquisition review board (JARB) and approved by commander, joint task force (CJTF).
- The PR&C is certified by the resource manager for funding and commitment.
- The PR&C and SOW are sent to contracting for execution and obligation.
- The goods or services are delivered to the customer.
- The vendor is paid by finance personnel.

6-2. The formal requirements for contracting relate primarily to the dollar limit of the acquisition. As the value of the goods or services increases, so do the required procedures. For example, acquisitions under $2,500 are generally accomplished by a unit FOO. Higher command approval is necessary when the amount exceeds $200,000.

6-3. There are reasons for the disparity in the processes. Some of these reasons are obvious, whereas other reasons are more obscure. The most obvious reason for the differences is that money always is in short supply when positioned against requirements. On the obscure side, fiscal laws governing disbursement carry harsh penalties for violations, including criminal penalties and loss of pay. Therefore, the requirements process has its own checks and balances that SF Soldiers must understand and follow.

6-4. Once the command has validated a high-dollar-value procedure—typically goods or services exceeding $200,000—there are three staff offices that work together and assist the unit in fulfilling their contracting requirements. These offices, commonly referred to as the "fiscal triad," are made up of the—
- Comptroller, who acquires, controls, and certifies funds IAW fiscal law.

Chapter 6

- Finance unit, who makes the payment.
- Contracting officer, who receives requirements and finishes the process of acquiring goods and services.

CONTRACTING PROCESS

6-5. There are three significant steps in the contracting process that SF Soldiers must understand. The three steps are—
- Identify and justify the requirements.
- Identify fiscal constraints and funding sources.
- Use the proper contracting authority and methods.

6-6. The SF unit must identify requirements to reduce the threat risk to the TACFAC. Potential requirements include items or services that make the TACFAC more secure but exceed on-hand resources. Crucial products during this process include the commander's justification memorandum, PR&C, SOW, and the IGCE. Strategic players include the FOO, PA officer, unit commander, installation property book officer (IPBO), JARB, and the Commander, Special Operations Command (COMSOC) (or his representative), for approval of PR&Cs.

IDENTIFY AND JUSTIFY REQUIREMENTS

6-7. Throughout the planning and construction processes, the SF unit should identify areas where changes or improvements may be made to the TACFAC layout, equipment, and procedures. Changes or improvements required to reduce the risk to people, mission, or property should be prioritized and forwarded up the chain of command. SF Soldiers should make full use of available analytical tools and procedures. For example, the SF unit may use Antiterrorism Planner—an Army Corps of Engineers tool for estimating damage—to analyze the projected effects of IEDs on tents, buildings, and personnel. Using the results of this analysis, the SF unit may determine requirements to reduce risk.

6-8. Requirements are categorized as either programmatic or procedural. Programmatic requirements include funds for bomb-detection equipment, additional security forces, and more military working dogs. Procedural requirements include additional personnel assigned to foot patrols, varying area spot checks by 10 minutes, and leaving the TACFAC via a different route.

6-9. Unless there is only a single possible source to fulfill a requirement, the SF unit should describe the requirement as generically as possible so that contracting personnel have more options when asked to fulfill it. Simply stated, the SF unit should identify *what* is needed and not *how* it is to be done. To do this, the SF unit should conduct a detailed analysis of the requirement and reduce it to its essence. The unit should be able to justify requirements using experience, proven operating procedures, and results from analytical tools. The SF unit writes the reasons for the requirements and the justification data, and then submits the documents through the appropriate staff to the unit commander for approval. The information is then developed into a JARB.

6-10. The main institutional tool used for identifying requirements (along with the justification, requested funding, and other data) is the Core Vulnerability Assessment Management Program (CVAMP). CVAMP is a web-enabled application resident within the (SIPRNET) Antiterrorism Enterprise Portal (ATEP) that captures the results of vulnerability assessments, prioritizes AO vulnerabilities, identifies deficiencies, and lists needed or completed corrective actions. CVAMP is located on the SIPRNET (http://www.atep.smil.mil). It is capable of storing classified data up to SECRET, and is accessible by SF commanders at every level. The CVAMP—
- Documents the SF commander's risk assessment decision for each vulnerability.
- Tracks the status of known vulnerabilities until addressed.
- Informs decision makers of the TACFAC ability to counter terrorist threats.

- Provides SF commanders a vehicle to identify requirements to the chain of command.
- Prioritizes resource requirements for request of Combating Terrorism Readiness Initiative Fund (CTRIF) support.

6-11. For the requirements review and contracting, the main document for identifying, approving, funding, and tracking a requirement within the Army is the PR&C. However, if manufactured supplies or services are needed, an SOW and an IGCE also are required. After the PR&C is drafted, the SF unit prepares the justification documentation, or JARB package. When justifying requirements, the following questions must be addressed:
- Why is the item or service needed?
- Why should the request receive priority?

6-12. The requirement may reduce the risk of death or injury; however, it must be explained sufficiently to pass the scrutiny of decision makers outside the SF unit. Whether or not a requirement is approved depends upon the threat, location, expected incident response, consequence-management actions (and their results), expected losses of personnel and equipment, and risk to the mission.

6-13. Analysis is often required to justify requirements; at times, this analysis may be intense. Analysis may include results from Antiterrorism Planner—a tool in the Joint Antiterrorism Program Manager's Guide (JAT Guide)—showing predicted blast damage to structures and personnel. The analysis may include the blast radius from a rocket or mortar, to include expected injuries. Multiple tools are available for predicting weapons effects on structures and personnel. Regardless of the technique, the SF unit must write a clear and convincing justification and present the requirements for approval. Verbal and written communication skills are crucial during this process.

6-14. There is a timeline for seeking funding. In 2004, during Operation IRAQI FREEDOM (OIF), the acquisition approval process followed a monthly cycle. Joint task force (JTF) staff officers held weekly meetings to consider and recommend disposition on pending theater requirements. The framework for these meetings was the JARB and the facilities utilization and support board (FUSB). The JARB and the FUSB were made up of intermediate representatives from all JTF-level staff sections. The JARB focused on all requirements exceeding $200,000. The FUSB was an engineer-oriented board that mandated uniform basing requirements across theater for the JARB to apply. Both of these boards were responsible for vetting requirements so the chief logistician and chief engineer could prioritize and approve requirements appropriately.

6-15. The JARB and FUSB were the final steps in the process. In the Iraq acquisition cycle, staff officers from the JARB and the FUSB would provide monthly briefings for the deputy commanding generals (DCGs) for support at each division regarding their duties and current actions. They would receive feedback about pending requirements. These meetings came to be known as the executive logistics review boards (ELRBs). After appropriate coordination, staff principles briefed senior leaders, who issued strategic guidance and provided priorities on theater-wide issues.

6-16. Because acquisitions exceeding $200,000 do not usually go to the comptroller or contracting without a JARB recommendation and an approval signature on the PR&C, SF Soldiers must understand how to develop an effective JARB package. Every JARB package must be properly staffed at the requesting unit and direct reporting unit before being submitted to the JARB.

Justification Memorandum

6-17. A request must have a justification memorandum that states clearly what the requirement is and what is requested to meet the requirement. The memorandum should describe how the requirement was determined and what options were considered to address the requirement. Units must address second-order and third-order effects of the request. For example, an SF unit submitting a request for trailers should also consider furniture, computers, and power hookups for the trailers. All items having to do with the trailer request should be included in a single request. Finally, the request must be signed by the appropriate level commander.

Analysis

6-18. The analysis must support the requirement. For example, if an SF unit requests large-capacity, COTS diesel generators, the unit must describe the MTOE and other COTS diesel generators assigned to the unit, how they are being used, and why the number or type of COTS diesel generators cannot meet the unit's mission requirements.

Staffing

6-19. As the SF unit's request is staffed through the chain of command, key personnel at each command level are provided the visibility of a much larger pool of resources. This provides an opportunity to fill the request internally or through other viable military options.

Funding Documentation

6-20. There are basically four types of funding documents that can be used to allocate funds. Determining which type of funding document is appropriate for a requirement may be difficult. The supporting resource management office and contracting personnel can help the SF unit determine which document is appropriate and how the document should be completed. The four basic funding documents are the PR&C, the military interdepartmental purchase request (MIPR), SOW, and IGCE.

Purchase Request and Commitment

6-21. The PR&C is used to buy something through local purchase or a new contract. For example, if the SF unit wants to lease a nontactical vehicle (NTV), a PR&C is used as the obligation document because it is a new purchase from a commercial source. The PR&C must list what the unit is trying to purchase and the cost of the purchase.

Military Interdepartmental Purchase Request

6-22. The MIPR is used to purchase something from a contract that already exists somewhere else in the DOD. For example, if an Army agency already has a contract in place for technical support, and the SF unit needs the same type support for the unit, funds are requested using a MIPR and allocated to the Army contracting agency that owns the contract. The contracting agency then makes the purchase under its contract.

Statement of Work

6-23. The SOW is used when trying to hire contractor support or labor. The SOW specifies the tasks to be performed, the standard to which the tasks must be performed, and what the SF unit (as the government sponsor) is going to provide. Without this information, contractors are unable to bid properly for the contract. Although writing a SOW normally is not a part of military training, nobody can define the requirement as well as the unit that needs the service or product. The most important thing for SF Soldiers to remember is that the supporting contracting personnel can provide assistance.

Independent Government Cost Estimate

6-24. The IGCE is the dollar-figure that the requester thinks is a fair and reasonable price for the product or service provided. SF units should attain estimates from a number of contractors to determine the "going price." An average of the figures provided should be included in the package as part of the USG estimate. Local contracting personnel can provide additional guidelines for an SF unit developing an IGCE.

FISCAL CONSTRAINTS AND FUNDING SOURCES

6-25. When seeking funding for a construction project at the TACFAC, the SF unit should obtain advice and assistance from the servicing command or SJA. These individuals should be able to help identify appropriate funding sources for the current and coming year, as well as any fiscal restraints upon the proposed project. Fiscal law, as it is known, is an ever-changing body of law that defines how money may be spent. More than any other area of law, SF commanders should expect and demand legal advice concerning fiscal law. This assistance is required by 10 USC, Sections 3037, 5148, 5046, and 8037.

Fiscal Constraints

6-26. Arguably the greatest challenge to a commander's career lies in his ability to comply with fiscal limitations. It is very important that SF leaders understand that fiscal law operates like no other form of law.

6-27. When seeking financial counsel most commanders ask, "Does anything prohibit me from doing this?" In the context of fiscal law, this is the wrong question to ask. The question that commanders should be asking is, "Are there any laws that authorize me to spend this money, in this way, and at the time I want to do this project?" This properly worded question addresses the three most important aspects of fiscal law: purpose, time, and amount. Money appropriated by Congress is never indefinite. Appropriated funds are designated for a certain purpose, for a limited period of time, and in a limited sum.

6-28. There are two primary types of appropriated monies immediately available to a CJTF during contingency operations: contingency operations money and humanitarian assistance money. Contingency operations money is a form of operations and maintenance (O&M) money that is appropriated for mission-essential operations in a particular contingency theater. Currently, in most taskers, the bulk of humanitarian assistance money is found in the commander's emergency response program (CERP). Proper mission-related expenses are those services or supplies needed to perform the mission or to support personnel performing the mission, and include the following:

- Services:
    - Cleaning services.
    - DFAC services.
    - Latrine services.
    - Utility services.
    - Power-generation services.
    - Air-conditioning services.
    - Tent services (to include set up, take down, and maintenance).
    - Trash services.
- Supplies:
    - Barriers.
    - Concertina wire.
    - Lumber.
    - Concrete.
    - Gravel (when used for maintenance and repair).
    - Batteries.
    - Radios.
    - Cellular telephones (however, air time is considered a service).
    - Generators.
    - Tents.
    - Paint.
    - Tools.
    - Spare parts.
    - Cots.
    - Furniture.
    - Air-conditioning units (when installed as an integral part of real property).
    - Medical supplies.
    - Oil.
    - Computers.

Chapter 6

6-29. Humanitarian assistance money in the CERP, unlike contingency operations money, can be used only by the CJTF in OIF to fund certain high-impact, high-visibility projects that benefit the HN people. SF units will find that contingency operations money funds most operations. Contingency operations money, however, carries with it the greatest restraints and pitfalls.

6-30. The SF unit's servicing SJA should be familiar with the types of money available and the purposes for which they can be used. The SJA should help ensure that the project is appropriately classified and that the proper rules are applied. The fiscal law questionnaire provided in Figure 6-1, pages 6-6 and 6-7, is designed to help SF Soldiers determine the appropriate classification and rules. The questionnaire highlights common issues, such as the $750,000 limitation on the use of O&M money for construction ($1.5 million for a requirement needed for life, health, or safety), the $3 million limit on repair, and the $250,000 limit on the purchase of individual items of supply. The questionnaire also identifies other concerns, such as items that are centrally managed (as with high-mobility multipurpose wheeled vehicles [HMMWVs] and Kevlar helmets) or projects or systems (such as equipment necessary for the operation of a computer network) that have been split apart to stay within applicable thresholds. The answers to the questions posed in the questionnaire help identify if Office of Military Affairs (OMA) or contingency operations money can be used, or whether other types of money (such as military construction [MILCON] or other procurement—Army [OPA]) funds must be obtained.

General Information
• What is to be acquired (for example, equipment, supplies, services, or construction)?
• Who will supply the product or service (for example, private contractor, DA, another DOD agency, or a non-DOD agency)?
• What legal process will be used to acquire the service, supplies, or equipment (for example, existing contract, new contract, Economy Act order, project order, General Services Administration [GSA] schedule, supply requisition, or logistics civil augmentation program [LOGCAP])?
• What contracting office or agency will process the acquisition?
• What kind of money will be spent (for example, Fiscal Year [FY] 2008 O&M)?
Supplies and Equipment
• What is the estimated date that the funds will be obligated?
• On what estimated date will the items be delivered and/or installed? (If the items will not be delivered/installed until the next FY, explain why.)
• Has delivery, testing, installation, and contractor temporary duty (TDY) costs been included in the total cost? (If not, explain why not.)
• Will any of the items be connected to equipment or systems already in the inventory? (If so, explain what function or capability the new equipment will add to the old.)
• For acquisitions costing more than $250,000:   ▪ If there is more than one unit or component, is each component a separate end item or system? Is it something that can stand alone or be used as part of a system?   ▪ If some of the units will be connected to form a functional system, describe the resulting system, its function, whether it will be temporary, how often it will be connected (whether it could later be reused for another purpose) and cost.   ▪ If the connected system of components described above has more than one function, describe the primary function. What are the secondary functions and why are they secondary?   ▪ Can some of the costs be attributed to installed building equipment (IBE)? If so, what are those costs? Attach a written opinion from a recognized expert that the components at issue will be IBE.   ▪ If acquiring software, describe the general function of the software (for example, software maintenance package, operating system software). What is the useful life of the software (for example, will the software require a major update within 2 years to meet changing requirements)?

Figure 6-1. Fiscal questionnaire

Supplies and Equipment (continued)

- Are any items centrally managed? This information can be obtained from the program manager for larger systems, such as aircraft, vehicles, weapons, and communications systems. Some centrally managed items have standard study numbers (SSNs), an 11-digit alphanumeric code maintained by Army Materiel Command (AMC).
- If the cost of each item is less than $250,000, O&M money can be used to purchase centrally-managed items provided the program manager provides a waiver stating that the requirement is recognized, yet cannot be supported in theater.
- For items priced at over $250,000, OPA funds must be used.

Services

- Describe all the services that the contractor will supply. Attach a copy of the relevant portions of the contract (if already in place) or a copy of the SOW.
- If funds are approved, when will funds be obligated?
- When will the services begin and end?
- If the services also involve acquisition of supplies or equipment, what estimated percentage of the total cost may be attributable to supplies and/or equipment?
- Describe why the services are nonpersonal services (contract employees are not supervised by government employees, are responsible for producing a product or result unrelated to how it is produced, and do not have an employer-employee relationship with the Army). The Federal Acquisition Regulation (FAR), Part 37, provides additional information.
- Describe why the services do not involve inherently governmental functions, such as supervising Soldiers, making policy decisions, or gathering intelligence. FAR, Part 7.5, provides additional information.
- Does the total cost include an estimated amount for increased costs when contract employees do not qualify as technical experts under the status-of-forces agreement (SOFA), or is there a clause in the contract shifting this burden to the contractor?

Construction and Real-Property Maintenance and Repair

- When will funds be obligated?
- When will actual performance begin and end?
- If the cost is more than $650,000, has a reasonable estimate (10% or more) been included for unexpected contingencies?
- If the total cost of the project exceeds $750,000 (MILCON threshold):
    - What is the project? Are structures permanent, one project, or many relocatable and reusable structures arguably defined as separate projects?
    - What portion of the project (if any) qualifies as construction? Construction includes building a new TACFAC, upgrades to improve the use of an existing facility, relocation of an existing facility, and conversion of a facility to different use. Local engineer or public works should be able to explain and provide an independent estimate.
    - Is the project necessary to protect the life, health, or safety of multinational forces? If yes, then up to $1.5 million of O&M funds can be spent.
    - What portion of the project (if any) qualifies as repair and/or maintenance? Maintenance is the preservation of an existing structure for day-to-day use. Repair is the restoration of an existing structure and the correcting of deficiencies. Separate the repair costs and the maintenance costs. The local engineer or public works should have the data. If the project is classed as repair or maintenance, the threshold for use of O&M money is up to $3 million.
    - If the project exceeds O&M thresholds; is vital to national security or the life, health, or safety of troops; and deferral to the next consideration by Congress of MILCON projects would be inconsistent with either national security or life, health, and safety, then the Secretary of the Army may use nonobligated MILCON money after giving notice to Congress and then waiting 15 days.

Figure 6-1. Fiscal questionnaire (continued)

Chapter 6

> ## Identifying and Interpreting Fiscal Regulations
>
> It should come as no surprise that SF TACFAC preparation (including relocatable buildings, sprung structures, tents, portable generators, and site preparation) were constant sources of fiscal scrutiny during OIF. Every Army engineer and SJA knows which regulations govern MILCON. However, the Assistant Chief of Staff for Installation Management (ACSIM) and the Army Comptroller regularly issued policy guidance on how to interpret and apply the principles of these regulations, as well as how to ask for available funds.
>
> For example, on 25 August 2004, the Army Comptroller issued a document titled *Procedures for Approval of O&M Construction Projects in Support of GWOT [Global War on Terror]*. This document required a centralized DD Form 1391 (Military Construction Project Data) process managed by USCENTCOM and the ACSIM. As a practical matter, this process made it more difficult to request MILCON funds and heightened the importance of proper project classification for the use of contingency operations money.
>
> Another recurring fiscal issue during OIF was not so much which "color" money to use—it became obvious that contingency operations was the only real source available for mission-essential construction needs. Rather, the real issue—and the area where the SJA's advice and assistance were best employed—was the proper classification of projects within fiscal (construction) limitations.

Funding Sources

6-31. Funded costs are costs that apply toward fiscal limitations and count against the thresholds. Examples of funded costs include—
- Materials, supplies, and services applicable to the project.
- Transportation costs for materials, supplies, and unit equipment.
- Installed capital equipment.
- Civilian labor costs.
- Overhead and support costs (for example, leasing and storing of equipment).
- Supervision, inspection, and overhead costs charged when the United States Army Corps of Engineers (USACE), the Naval Facilities Engineering Command, or the USAF serves as the design or construction agent.
- Travel and per diem costs for military and civilian personnel.
- O&M costs for government-owned equipment (for example, fuel and repair parts).
- Demolition and site-preparation costs.

6-32. Unfunded costs are costs that do not apply toward the fiscal limitations or count against thresholds. These costs are chargeable to appropriations other than those available to fund the project and are not reimbursed by appropriations available to fund the project. Examples of unfunded costs include—
- Military and civilian prisoner labor.
- Depreciation of government-owned equipment.
- Materials, supplies, and equipment obtained for the project on a nonreimbursable basis as excess distributions from another Service or USG agency.
- Licenses, permits, and other fees chargeable under—
    - State or local statutes.
    - SOFAs.
- Unfunded civilian fringe benefits.
- Contract or in-house planning and design costs.

Tactical Facility Funding

- Gifts from private parties.
- Donated labor and material contributed to the military construction project.

Note: Fiscal limitations do not count against repair and maintenance work, IAW Department of the Army Pamphlet (DA PAM) 420-11, *Project Definition and Work Classification*. The current limitation for these projects is $3 million. Projects which exceed this figure must be approved by Headquarters, Department of the Army (HQDA).

Operations and Maintenance

6-33. In their most basic form, O&M funds (to include contingency operations money) are used to conduct predeployment, deployment, and redeployment operations. These funds may be used to purchase fuel, barriers, forklifts, bulldozers, and sensors for early warning equipment. They also may be used to pay utility bills and hire custodial services. Generally, these funds can be used to make the SF TACFAC more secure. Services are authorized to use annual O&M funds for construction projects costing less than $750,000 ($1.5 million to correct a life-threatening condition or for new construction, and $3 million for maintenance and repair of existing facilities). This is a peacetime provision, applicable during contingencies and emergencies; however, a designation of a condition as "life-threatening" generally is considered a safety issue instead of a contingency operation emergency. During combat or designated contingency operations, O&M funds may be used to fund construction projects exceeding these thresholds under certain circumstances; however, the commander must consult with the SJA before making a determination to use O&M funds.

Military Construction

6-34. MILCON funds are obtained through a formal process using a DD Form 1391 and must be approved by Congress under applicable procedures. These funds are used to prepare the ground for construction; purchase bricks, mortar, concrete, and other construction materials; and pay construction labor, crane rental, and other expenses related to the construction of buildings, locks, dams, and roadways. Additionally, for the SF TACFAC, these funds may be used to build major buildings and structures, including concrete buildings, complex entry-control points, and a number of other phased projects.

Combating Terrorism Readiness Initiative Fund

6-35. The purpose of the CTRIF is to fund emergency (and emergent) high-priority combating terrorism (CbT) requirements in the year of execution. The funds provide a means for CCDRs to react to unforeseen requirements from changes in a terrorist threat, threat levels, or protective doctrine and standards. These funds also may be used by CCDRs to address unanticipated requirements identified during vulnerability assessments, tactical operations, and antiterrorism exercises. The CTRIF can be used to fund maintenance costs for CTRIF–funded items during the year of purchase (and the subsequent year, but only as a stop-gap measure). This allowance gives Services adequate time to program lifecycle costs if maintenance funds are not programmed and provided from the parent Service. CTRIF funds are not intended to subsidize ongoing projects, supplement budget shortfalls, or support routine activities—all of which are Service responsibilities.

Unfunded Requirements

6-36. Unfunded requirements (UFRs) are the needs for which there are no funds or insufficient funds during the current and following FYs. SF units should use the Planning, Programming, Budgeting, and Execution (PPBE) process to identify and justify requests to fulfill requirements. This is a long process compared to most activities—requested funds that are approved by higher HQ will not arrive for at least 2 years.

6-37. UFRs should be used to request funds for equipment and construction that are not available through O&M, CTRIF, or other local and Service funds. The servicing budget, program control, comptroller, or resource management office can help to identify deadlines and data-request formats.

Combatant Commander's Initiative Fund

6-38. The primary focus of the Combatant Commander's Initiative Fund (CCIF) is to support unforeseen contingency requirements critical to combatant command joint warfighting readiness and national security interests. The strongest candidates for approval are initiatives that support combatant command activities and functions, enhance interoperability, and yield a high benefit at a low cost. Protection of a TACFAC is a strong candidate because of its high benefit, low cost, and its support of combatant command activities and functions.

6-39. The CCIFs are not intended to subsidize ongoing projects, supplement budgetary shortfalls, or support expenses that normally are the responsibility of the parent Service. Initiatives considered by the Chairman of the Joint Chiefs of Staff (CJCS) in any FY are not eligible for resubmission or follow-on funding in subsequent FYs. Because all funds are in the O&M appropriation, all funding provided for approved projects must be obligated before the end of the FY for bona fide needs of that FY. Combatant command projects must be nominated for consideration by the CCDR or deputy CCDR. The CJCS is the final approval authority for CCIF requests. Chairman of the Joint Chiefs of Staff Instruction (CJCSI) 7401.01C, *Combatant Commander's Initiative Fund*, provides details regarding restrictions on the use of the CCIFs. Initiatives submitted for funding under this program must qualify within one of the following authorized activities:

- Joint exercises and force training.
- Contingencies and selected operations.
- Humanitarian and civic assistance (HCA).
- C2.
- Military education and training for military and related civilian personnel of foreign countries.
- Personnel expense of defense personnel for bilateral or regional cooperation programs.

6-40. There are a number of key questions the SF unit must answer when requesting CCIF support. These questions include the following:

- Is the request considered unforeseen or emergent and why?
- Is funding for this request to subsidize an ongoing project, supplement a budgetary shortfall, or support an expense that is normally the responsibility of the parent Service?
- Are there other funding sources for the request that fit the following and, if so, why are they not being used?
    - Initiatives already funded by CCDR's executive agent or components.
    - Normal Service operating costs (to include O&M costs).
    - Initiatives that have other available funding sources, such as annual HCA submissions or C2 projects.
- Does the request have an effect on the military operations in the post-9/11 world, strengthen joint warfighting capability, or aid in transforming the joint force?
- What is the realistic impact of failure to fund the effort?
- Is there a liaison or SME on the joint staff for this effort? If so, identify the POC.
- Is there a clear statement of need?
- Is there a detailed cost estimate that includes TDY requirements, contractual services, equipment purchases, and unit costs, rates, and descriptions of contractual vehicles to be used?

6-41. Requests for the CCIF should be made by letter. Submissions are not limited to one page. Adequate information is required for the joint staff to assess each initiative. Answer all the requests for data in the submission. Additional information may be found at the following Web site:

http://www.dtic.mil/cjcs_directives/cjcs/instructions.htm

Contracting Authority and Methods

6-42. The joint mission of the resource management and contracting offices is to fairly allocate scarce resources across a theater of operations. With command approval, resource management allocates funds to the contracting office, enabling it to obtain those supplies and services that the unit must have to perform its mission.

Funding Fragmentary Orders

During OIF, funding fragmentary orders (FRAGORDs) defined what services could be purchased for TACFACs, perhaps the largest continuing expense. This FRAGORD differentiated between TACFACs with less than 600 personnel and those with more than 600 personnel. This same FRAGORD also stated a preference for using the LOGCAP contract only as a last resort.

Most discussion in Iraq about contract type devolved into a discussion about whether or not to use LOGCAP. As the influence of the Joint Contracting Command increases in Iraq and Afghanistan and available funds decrease, the discussion of whether or not to use a cheaper method of contracting may resolve itself. At present, it remains an issue for every JARB and a consideration for every SF Soldier.

Contracting Authority

6-43. No individual can contract for the USG without proper authority. It is critical for SF Soldiers to understand that contracting authority is not derived from the same place or in the same manner as command authority. In the USCENTCOM AOR, for example, contracting falls under the responsibility of the ASAALT. Before a contract is signed, the person signing the contract should ensure that he has adequate authority—in writing, through a warrant of written appointment—to sign the contract.

6-44. Contracts entered into without authority must be ratified by the contracting chain of command. If a contract is not ratified, the unauthorized signer may be required to pay for the contract out of pocket. As such, if authority is unclear, the SF Soldier should contact the contracting office or the servicing SJA for guidance.

6-45. Given proper authority, contracts are the legal agreements between the USG and individuals or businesses for the delivery of services, products, and equipment. A contractor can be a single individual or a giant corporation. Contractors may be U.S. citizens, foreign nationals, or HN personnel.

Contracting Team

6-46. For the SF TACFAC, all contracts must incorporate protection and antiterrorism considerations. To effectively do this, the joint forward operating base (JFOB) needs a contracting team or needs to place the protection officer on the JFOB contracting team. The following key personnel should be on the team:
- Director of logistics division, director of contracting, or similar.
- Protection officer.
- Servicing SJA or legal officer.
- Contracting officer (KO) or FOO. These are the only people who can legally obligate the government to pay for materials and services.
- Contracting officer representative (COR) or contracting officer technical representative (COTR). These individuals cannot legally obligate the government to pay. These individuals often write SOWs and act as technical POCs for the KO, who may not be familiar with technology, materials, or tactics, techniques, and procedures (TTP).
- Task monitors. The COR or COTR also may serve as a task monitor. These personnel represent the unit initiating the contractual requirement.
- Comptroller, disbursing officer, or PA. These are the only people who can legally certify and deliver funds for payment to a vendor.

Chapter 6

Protection-Related Tasks in Contracting

6-47. Department of Defense Instruction (DODI) 2000.16, *DOD Antiterrorism Standards,* provides tasks to accomplish with regard to contracting while considering antiterrorism measures. These tasks are as follows:

- Implement a verification process, whether through background checks or other similar processes, that enables the USG to attest to the trustworthiness of DOD contractors and subcontractors (U.S. citizens and HN personnel), including those personnel having direct or indirect involvement in the delivery or provision of services related to mail, supplies, food, water, or other materiel and equipment intended for use by DOD personnel. This vetting of trustworthiness shall include husbanding agents and crews on contracted ships, planes, trains, and overland vehicles.
- Develop and implement site-specific risk-mitigation measures to maintain positive control of DOD contractor and subcontractor access to and within installations, sensitive facilities, and classified areas.
- Develop and implement site-specific risk-mitigation measures to screen contractor or subcontractor transportation conveyances for chemical, biological, radiological, nuclear, and high-yield explosives (CBRNE) hazards before entry into or adjacent to areas with DOD personnel and mission-essential assets.
- Ensure that contracts comply with the antiterrorism provisions of the Defense Federal Acquisition Regulation Supplement (DFARS).

Types of Contracts

6-48. There are various types of contracts; each has advantages and disadvantages. SF Soldiers should consult the servicing contracting office for the type of contract best suited to the requirements. The most common types of contracts are—

- Firm, fixed price (FFP).
- Cost-plus fixed fee (CPFF).
- Cost-plus award fee (CPAF).
- Time and materials (T&M).

6-49. Contracts can be modified using change orders. A change order adds, deletes, or changes tasks in the SOW. The change order provides flexibility when requirements change and the contract needs to be modified. Task orders usually are added to T&M contracts. Task orders often are placed on contracts when specific efforts (such as studies, analyses, and professional support services) are needed for a specified period of time. The task orders provide flexibility as needs change. Some examples of the products and services most likely acquired under contract are—

- Local transportation services.
- Lodging on local economy.
- Security services outside the TACFAC (if allowed by applicable law).
- Fire department services.
- Potable water delivery.
- Electricity production and delivery.
- Natural gas, propane, or butane delivery.
- Sewer services.
- Garbage collection and disposal.
- Medical and hazardous-material disposal.
- Mortuary services.
- POL.
- Meals.
- Tool kits (for example, for carpenters, plumbers, or electricians).

- Construction materials (for example, wood, brick, concrete, wallboard, nails, and screws).
- Construction services (for example, laborers, tradesmen, and craftsmen).
- Earth materials (for example, sand, gravel, and topsoil).
- Earth-moving equipment (for example, bulldozers, graders, scrapers, and dump trucks).

Contracting Methods

6-50. In most contingency environments there are two individuals that have authority to contract, and their methods differ. The first is the FOO, appointed by the unit. The second is a warranted KO, appointed by the principal assistant responsible for contracting (PARC) in theater. Both derive their authority from the contracting chain of command in theater. The primary difference in these individuals' authority is the amount of money for which they can contract and the formalities that must be followed for each.

Field Ordering Officer

6-51. The FOO generally uses a Standard Form (SF) 44 (U.S. Government Purchase Order Invoice Voucher) to contract and only has authority for contracts of $2,500 or less. This is the simplest form of contracting. In order for an FOO to be able to contract for goods or services, sufficient funds must be available from the PR&C. Additionally, all goods and services must be—
- Emergency or mission-critical.
- Not immediately available through military supply channels.
- Immediately available from local vendors (on-the-spot purchase).
- At or below $2,500 (no split requirements).
- Available as a one-time delivery (over-the-counter) with a one-time payment.
- Fair and reasonably priced (rotate vendors).

6-52. In order to properly contract and execute payment, there are several procedures that the FOO and PA must follow. These procedures are as follows:
- Receive the requirement from the unit commander.
- Ensure funds are available on a PR&C.
- Locate the vendor and inspect the item.
- Determine a fair and reasonable price for the item.
- Prepare an SF 44.
- Execute the purchase (PA pays the vendor).
- Receive an itemized receipt from the vendor.

Warranted Contracting Officer

6-53. Depending upon the terms of their warrant, KOs may have unlimited spending authority; however, they must follow strict procedures contained in the FAR and other applicable guidance. Figure 6-2, pages 6-14 and 6-15, provides a resourcing funds and contracting checklist. However, during contingency operations (and with the approval of either the PARC or the head of contracting activity), these requirements may be relaxed under certain specified circumstances. These circumstances require a written justification and approval describing unusual or compelling urgency. These documents must be reviewed by an SJA prior to approval by the appropriate level of the contracting chain of command. In Iraq, for example, the head of contracting activity possessed the statutory authority to make competition and FAR exceptions for acquisitions up to $75 million. Such exceptions by the head of contracting activity are not the norm, primarily because there are a number of other procurement flexibilities in a contingency environment designed to meet the needs of commanders. Whatever method is used, the process used by KOs follows a similar path.

- Define the exact requirement for the specific support, service, or product required.
- Identify alternatives to contracting for the service or support.
- Determine if local vendors or contractors accept the international merchant purchase authorization card (IMPAC).
- Determine if local vendors or contractors accept payment in local currency.
- Determine if local lodging providers accept payment in local currency or IMPAC.
- Determine if payments to local contractors will be in cash or by check.
- Identify who will provide security for the funds before and during disbursement if payments are made in cash.
- Identify who will provide security for disbursement personnel traveling to and from the TACFAC.
- Estimate the amount of cash needed to pay local vendors and contractors for products and services.
- Determine if the contracting task managers need translators.
- Identify how translators will be paid (cash, check, or electronic banking).
- Determine if escorts for local contractors will be needed.
- Develop and obtain approval for a method for letting local contractors into the TACFAC.
- Determine if enough real estate has been obtained through SOFAs or other means to provide adequate standoff protection from VBIEDs.
- Use the threat assessment and vulnerability assessment to determine contractual requirements.
- Identify the local capabilities for security and antiterrorism measures that must be considered if HN contractors are employed at the TACFAC.
- Designate a contractor entry point.
- Identify additional security measures that the contractor can provide during the service or support.
- Identify what the contractor can do to augment existing security arrangements.
- Determine if the contractor can properly vet the security clearance for all employees.
- Determine what measures of uncertainty still exist after vetting the contractor's employees.
- Determine if the unit considered asking for periods of support or service that are not routine or predictable in order to reduce risk of exposure.
- Determine if coordination with the contractor ensures a more predictable time period of coverage.
- Determine if local agencies can provide extra assistance if the contractor is unable to provide additional OPSEC or protection measures.
- Determine if the unit can add extra protection measures as needed.
- Where vetting cannot be achieved or additional protection measures cannot be contracted, identify what specific protection measures the unit can implement to narrow operational risk.
- Determine if the reason for the service or support requirement is really needed after all.
- Determine if the operational benefit of receiving the service or support outweighs the identified security shortfalls.
- Determine if the support tasks are mandated or can operational flexibility be employed to mitigate the overall risk.
- For support contracts, determine if the unit incorporated antiterrorism considerations into the contract, such as conducting background checks of all contractor and subcontractor employees.
- Establish a process for positively identifying all contractor and subcontractor employees and consider the use of the following:
  - Photo IDs with fingerprints.
  - Official IDs with fingerprints.
  - USG-issued IDs (only after background checks) with fingerprints.
  - Company-issued IDs (as last resort) with fingerprints.

Figure 6-2. Resourcing funds and contracting checklist

- Limited vehicle access.
- Daily personnel-access lists with photos and fingerprints.
- Daily vehicle-access list.
- Roster of all watercraft being used.
- Roster identifying all food and water sources being used.
* Determine if contractor understands, acknowledges, supports, and briefs appropriate company and subcontractor personnel on the antiterrorism measures to be implemented.
* Determine if the measures have been coordinated with and approved by appropriate local agencies.
* Determine if there is a reliance on local contracts for aircraft services and support.
* Identify the local security procedures at the selected airfields that contractors must follow.
* Determine if there is a means to identify and approve contractors to work at the airfields.

Figure 6-2. Resourcing funds and contracting checklist (continued)

6-54. During contingency operations, there are several simplified acquisition methods in the KO arsenal. These methods vary in complexity and approval requirements. SF units should consult with their SJA and KO to select the best method. In acquiring goods or services, the KO may use any of the following:

- Purchase orders. The KO offers to buy supplies, services, or construction. The contract occurs when written acceptance is received. Forms used during purchase orders include DD Form 1155 (Order for Supplies or Services), SF 33 (Solicitation, Offer, and Award), or SF 44.
- Government purchase card (GPC).
- Accommodation or purchase card checks.
- Blanket purchase agreements (BPAs). BPAs are advance agreements for future contracts that offer to repetitively use supplies or services. BPAs set the price, terms, and clauses to rapidly acquire items. BPAs do not require the USG to use the same contractor. The form used for a BPA is the DD Form 1155.
- Imprest funds. Imprest funds are petty cash funds established by the disbursing officer for cashiers. Imprest funds are specifically identified for certain use. The maximum amount advanced to units or cashiers is $10,000. Cash may be used to pay for micropurchases (limited to $2,500 each). These funds generally are used for immature theaters, and appropriated funds are used to reimburse the imprest funds. Each purchase handling imprest funds must be validated and authorized; cashiers cannot be ordering officers (separation of duty).
- Existing contracts. Contracts may exist with other Services or USG agencies that meet unit requirements. The KO can amend or use existing contracts.
- LOGCAP. Several contracts for support exist.
- Air Force Civil Augmentation Program (AFCAP). AFCAP is used primarily for civil engineering projects.
- Acquisition and cross-servicing agreements. The DOD has authority to acquire logistic support within NATO, United Nations (UN), and other nations approved by the Department of State (DOS).

This page intentionally left blank.

Chapter 7

Tactical Facility Relief in Place

JP 1-02 defines relief in place as "an operation in which, by direction of higher authority, all or part of a unit is replaced in an area by the incoming unit. The responsibilities of the replaced elements for the mission and the assigned zone of operations are transferred to the incoming unit. The incoming unit continues the operation as ordered." Transfer of authority (TOA) is the end state ceremony of the RIP passing authority and responsibility from the outgoing unit commander to the incoming unit commander. For this process to flow smoothly and with no discernable loss to mission continuity, the outgoing SF unit and commander must plan, coordinate, and synchronize closely with the incoming SF unit and commander. This process is also referred to as a mission handoff.

PROCEDURES

7-1. The overall authority for the RIP lies with the SF commander ordering the change. The responsibility for determining the details of the RIP lies with the incoming commander who will assume responsibility for the mission. The RIP process may affect the circumstances of mission continuation, ranging from command to operational climate.

7-2. The outgoing SFODA commander advises the incoming SFODA commander on the tentative RIP, his assumption of command, and the mission. If the outgoing SF commander's advice conflicts with the mission statement or the incoming commander's requirements, and the conflict cannot be resolved by the authority established for the incoming commander, their senior SF commander ordering the RIP will resolve any issues.

7-3. As a general rule, the senior SF commander ordering the RIP does not automatically place the outgoing SFODA under OPCON of the incoming SFODA. Although doing so would offer a clear and easily defined solution to establishing the incoming SF commander's authority, it is not the most effective control method, particularly if hostile contact occurs during the mission-handoff process.

7-4. If the incoming SFODA comes into direct fire contact with hostile forces during the RIP, the outgoing SFODA immediately notifies the HQ that ordered the RIP. If the incoming SFODA commander has not yet assumed responsibility, his SFODA immediately comes under OPCON of the outgoing SFODA and is absorbed into that SFODA position. If the outgoing SFODA commander already has passed responsibility to the incoming SFODA commander, the outgoing SFODA comes under OPCON of the incoming SFODA. The key for both incoming and outgoing commanders is to maintain operational continuity.

CONSIDERATIONS

7-5. During the incoming SFODA's assumption of control over a TACFAC from an outgoing SFODA, there are a number of operational considerations both commanders must resolve in order for the mission handoff to go smoothly. Both SFODAs and commanders spend the time available equally for the RIP. This means that initially the outgoing SFODA leads by example and the incoming SFODA observes. Halfway through the RIP they switch, and the incoming SFODA commander leads under the watchful eye of the

outgoing SFODA commander. Both SF commanders must consider the six critical METT-TC factors during the RIP process. Questions the incoming commander should have answers to include the following:
- How will the incoming and outgoing SFODAs infiltrate and exfiltrate the TACFAC?
- Will TACFAC movement be clandestine or overt (using military or civilian vehicles)?
- What are the TACFAC support requirements to conduct a successful RIP?
- Was a 100-percent TACFAC inventory conducted?
- Was all the stay-behind TACFAC equipment accounted for and signed for?
- Are there TACFAC shortage annex documents and requisitions for any missing items?

MISSION

7-6. The incoming SFODA commander must make a detailed study of the SFODA mission statement and understand the current mission tasks (to include implied mission tasks). The mission may require an SFODA with advanced skill sets, such as unique infiltration and exfiltration qualifications, Special Forces advanced urban combat (SFAUC) skills, or sniper skills. Knowing the mission, the commander's concept of the mission, the CCIRs, priority intelligence requirements (PIRs), and information requirements will facilitate the incoming SFODA commander's RIP process and provide a firm grasp of the mission. After an in-depth study of the current operations in the AO, the incoming SFODA commander completes the mission handoff in a manner that facilitates uninterrupted mission accomplishment. The RIP process is conducted in such a way that the enemy is denied an opportunity to gain any operational advantage.

ENEMY

7-7. The incoming SFODA commander must have the most current ISR data available on all enemy forces that may impact the mission. This data especially includes terrorists and terrorism-related incidents over the past several months. In addition to this intelligence provided to the incoming SFODA commander on a regular basis, the outgoing SFODA may provide an intelligence liaison element. OPSEC is critical to prevent the enemy from discovering the impending relief and possibly exploiting the RIP process. The outgoing SF commander provides continuous intelligence updates to the incoming SF commander. The outgoing SF commander also provides the PIRs and information requirements established for the initial mission, along with strategic, operational, and tactical information. The incoming SF commander must become completely familiar with the current PIRs and information requirements, as well as the PIRs and information requirements of any additional or newly assigned requirements.

TERRAIN AND WEATHER

7-8. The RIP may require both SFODAs to move dismounted through their AO. The outgoing SFODA commander plans and reconnoiters potential infiltration and exfiltration sites along established routes for both SFODAs. The selected routes in the AO should provide acceptable levels of cover and concealment. When possible, both SFODAs should plan movement during darkness or inclement weather.

TROOPS AND SUPPORT AVAILABLE

7-9. Information for the incoming SFODA commander about the friendly forces is as important as knowing the enemy situation. The SFODA must be familiar with the C2 structure it will work with on a daily basis. Additionally, the SFODA must be familiar with the identities, locations, operations, and capabilities of adjacent units. If possible, the incoming SFODA members should receive biographical data on the friendly forces. This will allow SFODA members to familiarize themselves with their counterparts prior to deployment. If support units are also to be relieved, these units should be relieved after the SFODAs they support.

7-10. The incoming SFODA plans and prepares for a quick and seamless transition in HN-counterpart relations. Any possibility of friction between the HN unit and incoming SFODA may cause the RIP to take longer than expected. To mitigate delays, the incoming and outgoing SFODAs should plan for a flexible time period for overlap before the mission handoff and to allow for actual face-to-face contact between the HN and all the SFODA participants.

TIME AVAILABLE

7-11. The depth and dispersion of friendly forces and the number of operations being conducted will dictate the amount of time required to conduct a RIP. There must be a transition overlap period for both SFODAs to allow the incoming SFODA to become familiar with the AO and to be introduced to and establish rapport with local civilians and their HN counterparts. However, the mission handoff must be completed as quickly as possible. A smooth RIP reduces time the enemy has to mount a strike before the incoming SFODA consolidates its position. However, the SFODAs should not sacrifice the quality of ongoing operations simply for extra time. The incoming SFODA needs to have enough time to observe daily tasks, training, and TTP, as well as to conduct debriefs on all their lessons learned.

CIVIL CONSIDERATIONS

7-12. The incoming SFODA must conduct an in-depth area study, with specific attention given to local HN problems. Popular or unpopular civilian support for U.S. activities taking place within the AO may directly influence changes in the mission statement. The outgoing SFODA must provide the incoming SFODA critical information that describes in detail all completed and ongoing local civic-action projects. The incoming SFODA should understand the functioning of the local and regional HN government, the status of any international agencies or NGOs involved, and their ability to influence the HN population in the AO.

TYPES OF RELIEF IN PLACE OR CLOSEOUT

7-13. Circumstances may occur during the RIP which require the TACFAC to be shut down. These conditions may range from a complete lack of support by the HN to a change in the U.S. strategic mission. When the SFODA is confronted with the task of a RIP, there are a number of possible COAs. The SFODA may be directed to perform one or more of the following:

- Abandon or destroy the TACFAC.
- Conduct a RIP of the TACFAC, transferring control to—
  - HN military or government.
  - UN or other international organization.
  - U.S. NGO.
  - USG or U.S. military.
  - Another SF unit.

ABANDON OR DESTROY THE TACTICAL FACILITY

7-14. SF units may be tasked to abandon or destroy a TACFAC. This process may be an emergency evacuation or a controlled evacuation. For emergency evacuations, the disposition of the TACFAC is conducted according to the overall evasion plan of action (EPA). Controlled evacuations usually take place when TACFACs have outlived their usefulness. In both scenarios, sensitive items are accounted for and either removed or destroyed to prevent their use by hostile forces. During both types of TACFAC evacuations, all Soldiers must understand where they need to go and what they need to do.

RELIEF IN PLACE TO THE HOST-NATION MILITARY OR GOVERNMENT

7-15. An SF unit turning over a TACFAC to the HN requires specific instructions from the DOS and DOD regarding the disposition of equipment within the compound. If the intent when building the facility was to eventually conduct a RIP to the HN at some point, clear guidance on how this process is to be accomplished should be given ahead of time. An important strategic and final coordination detail also must be considered—whether or not the U.S. will be reimbursed for construction costs.

Chapter 7

RELIEF IN PLACE TO THE UNITED NATIONS OR OTHER INTERNATIONAL ORGANIZATION

7-16. As with RIP to the HN military or government, turning over an SF TACFAC to the UN or other international organization is primarily a political decision. It must be done with clear guidance from the DOS and DOD.

RELIEF IN PLACE TO A U.S. NONGOVERNMENTAL ORGANIZATION

7-17. Depending upon the function of the establishment, SF TACFACs may be turned over to a U.S. NGO. Although the TACFAC is being handed over to another group of U.S. citizens, the transition to any NGO is a political decision that requires clear direction from the DOS.

RELIEF IN PLACE TO THE U.S. GOVERNMENT OR U.S. MILITARY

7-18. As circumstances in the region develop, it may be necessary to transfer control of the TACFAC to another USG agency or U.S. military organization. For example, a TACFAC set up to train and maintain counterdrug (CD) operations may eventually be transferred to the Drug Enforcement Administration (DEA).

RELIEF IN PLACE TO ANOTHER SPECIAL FORCES UNIT

7-19. Long-term operations may require one SF unit to conduct a RIP to another SF unit. Depending upon the length of the operation, the mission handoff process may occur dozens of times at the same TACFAC.

RELIEF IN PLACE ENVIRONMENTAL FACTORS

7-20. Once the SFODA enters the country where it plans to conduct operations, security precautions must be increased in order to successfully infiltrate and conduct a TACFAC RIP. For example, in the Iraq theater, SFODAs usually move into TACFACs clandestinely during hours of darkness. This helps to deter potential enemy attacks and hostile activity when the incoming and outgoing SFODAs are most vulnerable.

URBAN RELIEF IN PLACE

7-21. SF units typically move quickly, quietly, and with a minimal signature when en route to an urban TACFAC. Civilian vehicles common to the local area and civilian clothing (along with concealable body armor) help the unit to blend in with the local populace. SF units pay attention to the local culture, behaviors, and customs of the local populace and blend in to the HN environment. For example, although sunglasses are commonly worn in the United States, few Middle Eastern citizens ever wear them. The incoming SFODA should quiz the outgoing SFODA on what they have learned about local cultural customs and taboos. Questions the incoming SFODA should ask regarding the HN locals include the following:

- Are the HN locals friendly to the SF unit's presence in their AO?
- What is the threat level in the AO?
- Are any HN locals employed by the TACFAC?
- Is it common knowledge that U.S. forces occupy a TACFAC in the area?
- What are the safest routes locals use to travel?
- Are there any trustworthy local civilian or military leaders?

7-22. The incoming SFODA must be prepared to begin RIP procedures as soon as possible after arriving at the urban TACFAC. The outgoing SFODA normally has been deployed for a considerable length of time and wants to return to their home station.

7-23. The incoming 18C develops a checklist regarding the current state of the TACFAC and any issues that need to be addressed. Items on the checklist may include the following:

- Are there major outstanding contracts or concerns that must be corrected in the next 30 days?
- What daily, weekly, or monthly logistics are required to sustain the TACFAC?

- What special or SF-peculiar equipment or logistics are required, if any?
- How many HN personnel are staffed at the TACFAC?
- How long have the HN personnel worked there?
- What are the HN salaries and when do they get paid?
- What local contractors are used for sustainment and daily operations?
- Have these HN employees been vetted by qualified personnel?
- Are there any unfinished security or HN personnel issues?
- What daily functions or procedures occur that need to be addressed?
- Are groceries purchased locally? If so, who is designated to purchase them?
- Are there specific contractors used for infrastructure repair (such as water, plumbing, fuel, and electricity)?
- If the facility has generators for primary and backup sources of power, where does the fuel come from?

7-24. The incoming SFODA recognizes that the outgoing SFODA is the best source of information about their urban area—it is always better for the incoming unit to ask too many questions than not enough. During the RIP process, the outgoing SFODA must be prepared for the possibility that the incoming SFODA did not arrive with all of its supplies and equipment. When this happens, oversized cargo transportation must be coordinated for large SFODA equipment packages that cannot be carried by civilian vehicles. For example, an ISU-90 type container must be moved by a larger transport vehicle.

7-25. Along with proper inventory of the facility and equipment, the incoming 18B should review the security plan with the outgoing 18B. The existing security plan should cover all security measures in the event that hostile action occurs. Security areas of concern regarding the TACFAC include the following:
- Are there any immediate security measures that must be completed in the next 30 days?
- What are the weakest and strongest points of security at the TACFAC?
- How have the weak points been fixed or improved?
- Are the present physical barriers, obstacles, and wire adequate or are more needed?
- Are mobile communications between security personnel secure, or are they using nonsecure radios (such as the Motorola Talk-About) susceptible to radio direction finding and interception?
- What is the minimum safe distance from the outer barrier to the TACFAC in the event that a powerful VBIED detonates?
- Is there appropriate overhead cover available to protect against mortar attacks?
- Is there an early warning signaling system that alerts when an attack is imminent or over?
- What are the PACE evacuation plans for the SFODA and other personnel?

7-26. Both SFODAs should understand their responsibilities in the event of hostile activity during the urban RIP. Once the transition is complete, the incoming SFODA coordinates transportation for the outgoing SFODA to their point of departure. This move must be planned and coordinated in advance and thoroughly understood by both parties during the RIP process. The equipment load for the outgoing SFODA may be potentially larger than that of the incoming SFODA.

RURAL RELIEF IN PLACE

7-27. Reoccupying a TACFAC in a rural setting is similar in many ways to the RIP and occupation of an urban TACFAC. The planning, coordination, and synchronization duties of the 18C and 18B are almost identical in both urban and rural settings. The major differences between the urban and rural TACFAC RIP are when the TACFAC has a large contingent of HN forces or employees. The SF unit must ensure all—
- HN forces and civilians are current for pay.
- Existing contracts are complete or legally transferred.

7-28. All appropriate methods of infiltration are considered and war-gamed to determine which technique is best suited for the mission. The SFODA also should consider the following:
- Is the SFODA receiving HN support?
- Does the TACFAC positively impact the local area by hiring workers and using local vendors?
- Are indigenous forces dependent upon the TACFAC and SFODA, or is the TACFAC dependent on the local population?

7-29. Many urban TACFACs enjoy a high level of logistical sustainment that is absent in rural TACFACs. When operating in remote areas, SF planners must consider the following:
- Will the incoming SFODA be able to sustain present logistics requirements at the TACFAC?
- Are supplies delivered by parachute, helicopter, or by ground vehicles?
- Is there an available DZ and LZ or do they need to be built?
- What local supplies and vendors are available to the SFODA?
- How long can the SFODA sustain TACFAC operations independently without support from the JSOTF or SOTF?

7-30. When the rural RIP process is complete between the SFODAs, the incoming SFODA must give priority consideration to the exfiltration of the outgoing SFODA. Historically, SFODAs accumulate a great deal of equipment. How, when, and where the SFODA is exfiltrated are important planning considerations. The incoming SFODA commander must ensure that coordination and synchronization is made with the SOTF and JSOTF.

COMBINATION OF RURAL AND URBAN RELIEF IN PLACE

7-31. Based on changing requirements, the final status of any TACFAC may change and revert to a combination of military and civilian organizations. Regardless of the new TACFAC owner, the RIP process must be a priority COA early in the MDMP.

7-32. The end SFODA occupation does not necessarily indicate the TACFAC is no longer needed. When the initial mission authorized construction of the TACFAC, it contained at least basic instructions for the proper disposition of the TACFAC once the SFODA is scheduled to return. If no such instructions exist, it is incumbent on the SF TACFAC commander and SFODA to learn the ultimate fate of the TACFAC so that appropriate measures may be taken.

CLOSEOUT ASSISTANCE

7-33. Operational planning is continuous, and the ultimate goal for the TACFAC is revised to reflect the existing political and military climate. For example, the initial EPA for a TACFAC may include vehicles and personnel support from the HN. This plan can (and should) be reviewed and revised as necessary, especially when a U.S. carrier task force enters the region.

7-34. CA teams may be provided to assist the SF unit in a TACFAC closeout. This option is particularly effective when the facility is to be turned over to an NGO and no suitable provisional government exists to assume control. PSYOP units may also be used to assist in closeout by establishing programs to explain to the local HN population the procedures and rationale for the particular disposition of the TACFAC and to prepare them emotionally for the change.

> *Note:* A fully functioning active TACFAC that employs HN locals for its support will create a large gap in the local HN economy with the departure of U.S. SF. Plans for this transition should be incorporated in the final disposition of the TACFAC to reduce the economic and political impact on the local population.

FINAL U.S. CLOSEOUT

7-35. Final U.S. TACFAC closeout includes a number of personnel, logistical, and financial considerations. These considerations include—
- Assembly of HN forces and employees of the facility into assembly areas.
- Completion of administrative records of all personnel at the TACFAC, to include a complete inventory of arms, equipment, and sensitive items.
- Settlement of pay, allowances, and benefits to HN personnel employed at the facility.
- Fair settlement of claims against the HN government or the United States.
- Recommendations for awards and decorations submitted for deserving HN personnel who supported the TACFAC.
- Rehabilitation and employment of HN personnel who supported the TACFAC.

7-36. The U.S. commander in charge during the final disposition of the TACFAC is responsible for direction, advice, and final guidance to all remaining U.S. and HN personnel. Additionally, this commander must coordinate and supervise any closeout operations that fall within his jurisdiction during this phase of operations.

This page intentionally left blank.

Appendix A
Area Study

Figure A-1, pages A-1 through A-5, provides an outline format for an SF area study. This format is UNCLASSIFIED until filled in with operational data. Then, the classification is determined by the appropriate classification authority.

(CLASSIFICATION)

Copy_____ of _____Copies   _____ SFG(A)

Location

Date

AREA STUDY OF JOINT SPECIAL OPERATIONS AREA

I. **PURPOSE AND LIMITING FACTORS.**
   A. Purpose. Delineate the area being studied.
   B. Mission. State the mission that the area study supports.
   C. Limiting factors. Identify factors that limit the completeness or accuracy of the area study.

II. **GEOGRAPHY, HYDROGRAPHY, AND CLIMATE.** Divide the operational area into its various definable subdivisions and analyze each subdivision as outlined below.
   A. Areas and dimensions.
   B. Strategic locations:
      1. Neighboring countries and boundaries.
      2. Natural defenses, including frontiers.
      3. Points of entry and strategic routes.
   C. Climate. Note variations from the norm and the months in which they occur. Note any extremes in climate that could affect operations, particularly the following:
      1. Temperature.
      2. Rainfall and snow.
      3. Wind and visibility.
      4. Light data. Include times for begin morning nautical twilight (BMNT), end evening nautical twilight (EENT), sunrise, sunset, moonrise, and moonset.
      5. Seasonal effect of the weather on terrain and visibility.
   D. Relief:
      1. General direction of mountain ranges or ridgelines and whether hills and ridges are dissected.
      2. General degree of slope.
      3. Characteristics of valleys and plains.

(CLASSIFICATION)

Figure A-1. Special Forces area study

Appendix A

(CLASSIFICATION)

    4. Natural routes for (and natural obstacles to) cross-country movement.
    5. Location of area suitable for guerrilla bases, units, and other installations.
    6. Potential LZs, DZs, and other reception sites.

E. Land use. Note any peculiarities, particularly the following:
    1. Former heavily forested land areas subjected to widespread cutting or dissected bypaths and roads. Also note pastureland or wasteland that has been restored.
    2. Former wasteland pastureland that has been resettled and cultivated and is now being farmed. Also note former rural countryside that has been depopulated and allowed to return to wasteland.
    3. Former swampland or marshland that has been drained, former desert or wasteland now irrigated and cultivated, and lakes created by dams.

F. Drainage. Note the general drainage pattern, particularly the following:
    1. Main rivers and their direction of flow.
    2. Characteristics of rivers and streams. Note widths, currents, banks, depths, bottom types, and obstacles.
    3. Seasonal variations. Note dry beds and flash floods.
    4. Large lakes or areas with many ponds or swamps. Note potential LZs for amphibious aircraft.

G. Coast. Examine for infiltration, exfiltration, and resupply points, noting the following:
    1. Tides, waves, winds, and current.
    2. Beach footing and covered exit routes.
    3. Quiet coves and shallow inlets and estuaries.

H. Geological basics. Identify types of soil and rock formations, and potential LZs for light aircraft.

I. Forests and other vegetation:
    1. Natural or cultivated.
    2. Types, characteristics, and significant variations from the norm at different elevations.
    3. Cover and concealment. Include density and seasonal variations.
    4. Water.

J. Subsistence:
    1. Seasonal or year-round.
    2. Cultivated. Include vegetables, grains, fruits, and nuts.
    3. Natural. Include berries, fruits, nuts, and herbs.
    4. Wildlife. Include animals, fish, and fowl.

III. **POLITICAL CHARACTERISTICS.** Identify friendly and hostile political powers and analyze their capabilities, intentions, and activities that influence mission execution, particularly the following:

A. Hostile power:
    1. Number and status of non-national personnel.
    2. Influence, organization, and mechanisms of control.

B. National government:
    1. Government, international political orientation, and degree of popular support.
    2. Identifiable segments of the population with varying attitudes and probable behavior toward the United States, its allies, and the hostile power.

(CLASSIFICATION)

Figure A-1. Special Forces area study (continued)

(CLASSIFICATION)

    3. National historic background.
    4. Foreign dependence or allies.
    5. National capital and significant political, military, and economic concentrations.
  C. Political parties:
    1. Leadership and organizational structure.
    2. Nationalistic origin and foreign ties (if single dominant party exists).
    3. Major legal parties with their policies and goals.
    4. Illegal or underground parties and their policies and goals.
    5. Violent opposition factions within major political organizations.
  D. Control and restrictions:
    1. Documentation.
    2. Rationing.
    3. Travel and movement restrictions.
    4. Blackouts and curfews.
    5. Political restrictions.
    6. Religious restrictions.

IV. **ECONOMIC CHARACTERISTICS.** Identify economic factors that influence mission execution, particularly the following:

  A. Technological standards.
  B. Natural resources and degree of self-sufficiency.
  C. Financial structure and dependence on foreign aid.
  D. Monetary system:
    1. Value of money and rate of inflation.
    2. Wage scales.
    3. Currency controls.
  E. Black market activities. Note the extent and effect of those activities.
  F. Agriculture and domestic food supply.
  G. Industry and level of production.
  H. Manufacture and demand for consumer goods.
  I. Foreign and domestic trade and facilities.
  J. Fuels and power.
  K. Telecommunications adequacy (by U.S. standards).
  L. Transportation adequacy (by U.S. standards):
    1. Railroads.
    2. Highways.
    3. Waterways.
    4. Commercial air installations.
  M. Industry, utilities, agriculture, and transportation. Note the control and operation of each.

(CLASSIFICATION)

Figure A-1. Special Forces area study (continued)

Appendix A

(CLASSIFICATION)

V. **CIVIL POPULACE.** Pay particular attention to those inhabitants in the AO that have peculiarities and that vary considerably from the normal national way of life. Note the following:
    A. Total population and density.
    B. Basic racial stock and physical characteristics:
        1. Types, features, dress, habits.
        2. Significant variations from the norm.
    C. Ethnic and religious groups. Analyze these groups to determine if they are of sufficient size, cohesion, and power to constitute a dissident minority of some consequence. GTA 41-01-005, *Religious Factors Analysis*, provides more information about religious groups. Note the following:
        1. Location or concentration.
        2. Basis for discontent and motivation for change.
        3. Opposition to the majority or the political regime.
        4. Any external or foreign ties of significance.
    D. Attitudes. Determine the attitudes of the populace toward the existing regime or hostile power, the resistance movement, and the United States and its allies.
    E. Division between urban, rural, or nomadic groups. Note the following:
        1. Large cities and population centers.
        2. Rural settlement patterns.
        3. Area and movement patterns of nomads.
    F. Standard of living and cultural (educational) levels. Note the following:
        1. Extremes away from the national average.
        2. Class structure. Identify degree of established social stratification and percentage of populace in each class.
    G. Health and medical standards. Note the following:
        1. General health and well-being.
        2. Common diseases.
        3. Standard of public health.
        4. Medical facilities and personnel.
        5. Potable water supply.
        6. Sufficiency of medical supplies and equipment.
    H. Traditions and customs, particularly taboos. Note wherever traditions and customs are so strong and established that they may influence an individual's actions or attitude, even during a war situation.

VI. **MILITARY AND PARAMILITARY FORCES.** Identify friendly and hostile conventional military forces (army, navy, and air force) and internal security forces (including border guards) that can influence mission execution using the following subdivisions:
    A. Morale, discipline, and political reliability.
    B. Personnel strength.
    C. Organization and basic deployment.
    D. Uniforms and unit designations.
    E. Ordinary and special insignia.

(CLASSIFICATION)

Figure A-1. Special Forces area study (continued)

(CLASSIFICATION)

- F. Overall control mechanism.
- G. Chain of command and communication.
- H. Leadership. Note officer and NCO corps.
- I. Non-national surveillance and control over security forces.
- J. Training and doctrine.
- K. Tactics. Note seasonal and terrain variations.
- L. Equipment, transportation, and degree of mobility.
- M. Logistics.
- N. Effectiveness. Note any unusual capabilities or weaknesses.
- O. Vulnerabilities in the internal security system.
- P. Past and current reprisal actions.
- Q. Use and effectiveness of informers.
- R. Influence on and relations with the local populace.
- S. Psychological vulnerabilities.
- T. Recent and current unit activities.
- U. Reconnaissance activities and capabilities. Pay particular attention to reconnaissance units, special troops (airborne, mountain, or Ranger-type units), rotary-wing or vertical-lift aviation units, CI units, and units having a mass CBRNE delivery capability.
- V. Guard posts and wartime security coverage. Note the location of all known guard posts or expected wartime security coverage along the main LOCs (railroads, highways, and telecommunications lines) and along electrical power and POL lines.

VII. **EFFECTS OF CHARACTERISTICS.** State conclusions reached through analysis of the facts developed in previous paragraphs:
- A. Effect on hostile COAs.
- B. Effect on friendly COAs.

(CLASSIFICATION)

Figure A-1. Special Forces area study (continued)

This page intentionally left blank.

Appendix B
Site Survey

The site survey checklist is a planning tool used by an SF site survey team. When filled out, it helps answer S-2, S-3, and S-4 TACFAC questions identified during the survey or during the early stages of operations. The site survey checklist shown in Figure B-1, pages B-1 through B-3, is a guide; it is not meant to be all-inclusive. The checklist can be modified as needed. This format is UNCLASSIFIED until filled in with operational data. Once filled in, the classification level is determined by the appropriate classification authority.

Security Assistance Organization
S-2
I. Intelligence briefing.
II. Threat briefing.
III. Maps and imagery of the area.
IV. Weather forecast data.
V. Restricted and off-limits areas.
VI. Local populace (religion, attitudes, customs, taboos, and dangers).
S-3

I. Initial coordination.
   A. Tentative TACFAC construction plans.
   B. Tentative training plans.
   C. Aviation support tentatively available (hours and type of aircraft).
   D. HN plans (tentative).
   E. Problem areas.
   F. EPA-related directives, guidance, plans, or orders.

II. POC, phone number list, communications requirements, and systems used.
S-4
I. Transportation requirements.
II. Special equipment requirements.
III. Other support requirements.
IV. TACFAC construction equipment and supply requirements.

Figure B-1. Site survey checklist

Appendix B

Host Unit
Commander
I. Construction and training plan.
II. Current training status.
III. Units available for training.
IV. C2.
V. Additional training desires.
VI. Unit policies.
S-2
I. Local civilians.
II. Protection and security policies and problems.
II. Populace control requirements (identification cards or passes).
S-3
I. Training plan.

II. Support available.
   A. Ammunition.
   B. Weapons.
   C. Vehicles.
   D. Aircraft or air items.
   E. Facilities and equipment:
      1. Training areas.
      2. Classrooms.
      3. Ranges.
      4. Training aids.
      5. Special equipment.

III. Unit equipment.
IV. LZs, DZs, and HLZs in the area.
V. Maps and other imagery.
VI. Rations for field training.
VII. Daily training schedules and status reports.
VIII. POC for training problems.
IX. Holidays, customs, taboos, and unit requirements that may interfere with training.

Figure B-1. Site survey checklist (continued)

S-3 (continued)
X. Medical and dental support. XI. Communications capabilities. XII. HN activities. XIII. Other.
S-4

I. TACFAC construction.
   A. Barracks.
   B. Drinking water.
   C. DFAC.
   D. Secure storage areas.
   E. Electrical power supply.
II. Fuel supply.
III. Rations.
IV. Transportation.
V. Lumber and materials for construction and training aids.
VI. Special equipment.
VII. Ammunition.
VIII. Availability of construction resources, equipment, tools, and supplies.

Figure B-1. Site survey checklist (continued)

This page intentionally left blank.

Appendix C
Area Assessment

In order to plan, train, and conduct SO, SFODA commanders require specific and current data about their AO. This data, when collected and confirmed, is called an area assessment. This format is UNCLASSIFIED until filled in with operational data. Then, the classification is determined by the appropriate classification authority.

GENERAL

C-1. The area assessment is a valuable tool used for the immediate and continuing collection of information started after infiltration into an AO. Characteristically it—

- Confirms, corrects, or refutes previous intelligence of the AO acquired as a result of the area study and other data sources prior to the infiltration.
- Forms the basis for tactical, operational, and logistical planning for the AO.
- Includes information on METT-TC and terrain analysis.
- Includes information on the different religious and ethnic tribal elements of the indigenous population in the AO.

C-2. The results of the area assessment are transmitted to the SOTF when there is marked deviation from previous intelligence. The SOTF prescribes in appropriate SOPs and annexes those items that are reported. The three types of area assessments are initial, principal, and preventative medicine.

INITIAL AREA ASSESSMENT

C-3. Initial area assessments include those items deemed essential to the SFODA immediately following infiltration. These requirements must be satisfied as quickly as possible after the SFODA arrives in the AO and should, at a minimum, include those items listed in Figure C-1.

I. Location and orientation.

II. SFODA physical condition.

III. Overall security issues:
   A. Immediate area.
   B. Attitude of the local populace toward HN government and SF presence.
   C. Local enemy situation.

IV. Status and type of the local insurgent elements.

Figure C-1. Initial area assessment

PRINCIPAL AREA ASSESSMENT

C-4. The principal area assessment is an ongoing assessment that includes those collection efforts that support the continued planning and conduct of operations. The principal area assessment (detailed in Figure C-2, pages C-2 through C-4) forms the basis for all subsequent SFODA activities in the AO.

Appendix C

(CLASSIFICATION)

I. Enemy:
   A. Morale and motivation.
   B. Composition, identification, and strength.
   C. Organization, armament, and equipment.
   D. Degree of training, morale, and combat effectiveness.
   E. Operations.
      1. Recent and current activities of the unit.
      2. Insurgent activities and capabilities. Give particular attention to IEDs, VBIEDs, suicide bombers, reconnaissance elements, rotary-wing or vertical-lift aviation units, and units having a mass CBRNE delivery capability.
   F. Unit AO.
   G. Daily routine of the units.
   H. Logistical support, to include—
      1. Installations and facilities.
      2. Supply routes.
      3. Methods of troop movement.
   I. Past and present reprisal actions.

II. Security and police units:
   A. Dependability and reliability to the existing regime or the occupying power.
   B. Morale and motivation.
   C. Composition, identification, and strength.
   D. Organization, armament, and equipment.
   E. Degree of training, morale, and efficiency.
   F. Utilization and effectiveness of confidential informants.
   G. Influence on and relations with the local populace.
   H. Active and passive security measures in place for public utilities and government installations.

III. Civil government:
   A. Controls and restrictions, such as—
      1. Documentation.
      2. Rationing.
      3. Travel and movement restrictions.
      4. Blackouts and curfews.
   B. Current value of money and wage scales.
   C. The extent and effect of the black market.
   D. Political restrictions.
   E. Religious restrictions.
   F. Control and operation of industry, utilities, agriculture, and transportation.

(CLASSIFICATION)

Figure C-2. Principal area assessment

Area Assessment

(CLASSIFICATION)

IV. Civilian population:
    A. Attitude toward the existing regime or occupying power.
    B. Attitude toward the resistance movement.
    C. Reaction to U.S. support of the resistance.
    D. Reaction to enemy activities within the country, specifically that portion which is included in guerrilla warfare operational areas.
    E. General health and well-being.

V. Potential targets (for each, consider population reaction):
    A. Railroads.
    B. Telecommunications.
    C. POL.
    D. Electric power.
    E. Military storage and supply.
    F. Military HQ and installations.
    G. Radar and electronic devices.
    H. Highways.
    I. Inland waterways and canals.
    J. Seaports.
    K. Natural and synthetic gas lines.
    L. Industrial plants.
    M. Key personalities.

VI. Weather:
    A. Precipitation, cloud cover, temperature, visibility, and seasonal changes.
    B. Wind speed and direction.
    C. Light data, including BMNT, EENT, sunrise, sunset, moonrise, and moonset.

VII. Terrain:
    A. Location of areas suitable for guerrilla bases, units, and other installations.
    B. Potential LZs, DZs, and other reception sites.
    C. Routes suitable for—
        1. Guerrilla forces.
        2. Enemy forces.
    D. Barriers to movement.
    E. Seasonal effect of the weather on terrain and visibility.

VIII. Resistance movement:
    A. Guerrillas:
        1. Disposition, strength, and composition.
        2. Organization, armament, and equipment.

(CLASSIFICATION)

Figure C-2. Principal area assessment (continued)

(CLASSIFICATION)

    3. Status of training, morale, and combat effectiveness.
    4. Operations to date.
    5. Cooperation and coordination between various existing groups.
    6. General attitude towards the United States, the enemy, and various elements of the civilian population.
    7. Motivation of the various groups and their receptivity.
    8. Caliber of senior and subordinate leadership.
    9. Health of the guerrillas.

  B. Auxiliaries or the underground:
    1. Disposition, strength, and degree of organization.
    2. Morale, general effectiveness, and type of support.
    3. Motivation and reliability.
    4. Responsiveness to guerrilla or resistance leaders.
    5. General attitude towards the United States, the enemy, and various guerrilla groups.

IX. **Logistics capability of the area:**
  A. Availability of food stocks and water, to include any restrictions for reasons of health.
  B. Agriculture capability.
  C. Type and availability of transportation of all categories.
  D. Types and location of civilian services available for manufacture and repair of equipment and clothing.
  E. Supplies locally available, to include type and amount.
  F. Medical facilities, to include personnel, medical supplies, and equipment.
  G. Enemy supply sources accessible to the resistance.

(CLASSIFICATION)

Figure C-2. Principal area assessment (continued)

PREVENTATIVE MEDICINE AREA ASSESSMENT

C-5. Preventive medicine services encompass those activities that are geared toward preventing or reducing the incidence of disease and injury. Because SF Soldiers are deployed into areas where the presence of endemic and epidemic diseases is high, where basic sanitation facilities and practices may not exist, where preventative medicine assets may not be readily available, and where environmental conditions may be adverse, these Soldiers are at a higher risk of injury and illness. The preventative medicine area assessment provided in Figure C-3, pages C-5 through C-9, is a useful tool for SFODAs (particularly 18Ds) deploying to an unfamiliar AO.

(CLASSIFICATION)

I. **Climatology and topography:**
   A. Cold weather:
      1. How long is the cold weather season?
      2. What is the temperature range?
      3. Is the weather cold enough to put emphasis on causes, treatments, and prevention of cold-weather injuries?
      4. Is it dry cold or wet cold?
   B. Hot weather:
      1. How long is the hot-weather season?
      2. What is the temperature range?
      3. Is the weather hot enough to put emphasis on causes, treatment, and prevention of heat injury?
      4. Is acclimatization necessary? If so, how many days?
      5. Describe humidity.
   C. Terrain:
      1. What types of terrain are present?
      2. What is the range in elevation?
      3. Are there swamps present?
      4. Are rivers present?
      5. Describe terrain over where the unit will operate and where the camp is located.
      6. How does the terrain affect evacuation and medical resupply?

II. **Indigenous personnel:**
   A. Physical characteristics:
      1. Give names of tribes or groups the unit will associate with.
      2. Describe indigenous personnel in terms of height, body build, and color and texture of skin and hair. Provide photographs if available.
      3. Describe the endurance, ability to carry loads, and other physical capabilities of the indigenous personnel.
   B. Dress:
      1. Describe clothing, principal ornamentation, and footwear.
      2. What symbolism is attached to various articles of clothing and jewelry, if any?
      3. Are amulets worn and what do they symbolize? Provide photographs if available.
   C. Attitudes:
      1. What taboos and other psychological attributes are present in the society?
      2. What are attitudes toward birth, puberty, marriage (for example, monogamy or polygamy), aging, sickness, and death? Describe any rituals associated with these events.
      3. Describe attitudes toward doctors and Western medicine.
      4. Describe rights and practices by witch doctors during illness. What do these rites symbolize? Does the practitioner use Western medicines?

(CLASSIFICATION)

Figure C-3. Preventative medicine area assessment

Appendix C

(CLASSIFICATION)

5. Do indigenous personnel respond to the events (for example, fear, happiness, anger, or sadness) in the same manner an American would?

D. Housing:
  1. Describe the physical layout of the community. Provide aerial photographs, diagrams, and notes.
  2. Describe construction techniques and materials used. Give the layout of a typical house (for example, sleeping areas and cooking areas).
  3. Describe infestation with ectoparasites and rodents.
  4. Describe general cleanliness.
  5. List the number of people inhabiting a typical dwelling.

E. Food:
  1. Describe the ordinary diet.
  2. Is food cultivated for consumption? Which foods?
  3. Describe agricultural practices (for example, clear-cutting and permanent farms).
  4. Do hunting, fishing, and gathering wild crops contribute significantly to the diet? What wild vegetables are consumed?
  5. How is food prepared? What foods are pickled, smoked, cooked, or eaten raw? How are foods preserved?
  6. At what age are children weaned? What diet is provided infants after weaning occurs?
  7. Which members of the family are given preference at the table?
  8. Are there food-related taboos? Is milk denied children, infants, and pregnant or lactating women? Why?
  9. How do the seasons influence diet? Are there famines? Does migration in search of food occur?
  10. What foods provided by U.S. personnel do indigenous personnel prefer or reject?
  11. What cash crops are raised?

III. Water supply, urban:
  A. Is water plentiful?
  B. Is water treated?
  C. What kind of water treatment plants are used (if any)?

IV. Water supply, rural:
  A. List the numbers and types of rural water supplies.
    1. What are the sources (for example, rivers, springs, or wells)?
    2. Is the same water used for bathing, washing, and drinking?
    3. How far is the water source from the community?
    4. Is water plentiful or scarce? Give seasonal relationships to water availability. Is water rationed by indigenous personnel?
  B. What treatment is given to water in rural areas?
    1. Is water boiled?
    2. Is water filtered or subjected to other purification process before consumption?

(CLASSIFICATION)

Figure C-3. Preventative medicine area assessment (continued)

(CLASSIFICATION)

    3. What is the attitude of indigenous personnel toward the standard U.S. purification methods?
    4. Is the untreated water safe for bathing?

V. **Sewage disposal (when applicable):**
    A. What are the types and locations of sewage treatment plants?
    B. Do urban areas have combined or separate sewage systems?
    C. What are the locations of outfalls and conditions of receiving streams before and after receiving affluent? State whether or not affluent is chlorinated and, if so, how much.
    D. What types of sewage disposal is used in those areas not connected to a sewer system? Give approximate numbers of—
        1. Septic tanks.
        2. Cesspools and leaching pits.
        3. Sanitary privies.
    E. In the community, what system is used for disposal of human excrement, offal, and dead animals or humans?
        1. Is excrement collected for use as a fertilizer? How is it done?
        2. What scavengers commonly assist in the process of disposal (for example, animals or birds)?
        3. What is the relationship of disposal sites to the watering sites?
        4. What is the attitude of indigenous personnel to standard U.S. methods, such as the use of latrines?

VI. **Epidemiology.** Describe specific diseases present among the guerrillas, dependents, and their animals. Relate the occurrence of more than one similar illness at a given time to a common source or event. Are there occurrences of—
    A. Respiratory disease?
    B. Arthropod-borne disease?
    C. Intestinal disease?
    D. Venereal disease?
    E. Miscellaneous diseases, such as tetanus, scabies, and dermatophytosis?

VII. **Domestic animals:**
    A. What domestic animals are present?
        1. Are they raised for food, labor, or other purposes?
        2. Of what importance is the herd or flock economically?
        3. Is any attempt made to breed animals selectively?
    B. Describe the normal forage.
        1. Do owners supplement the food supply? What supplements are given?
        2. Are animals penned or allowed to roam?
        3. Where are the animals housed?
    C. Is there any religious symbolism or taboo associated with animals? Are animals sacrificed for religious purposes?
    D. Describe in detail the symptoms and signs of any diseased animals. List the species involved.

(CLASSIFICATION)

Figure C-3. Preventative medicine area assessment (continued)

Appendix C

(CLASSIFICATION)

E. If disease is epidemic, describe the epidemic.
  1. What percentage of the herd or flock demonstrated illness?
  2. Did all the illnesses occur spontaneously or over a period of time?
  3. Was more than one herd or flock affected at the same time and location?
  4. Did the epidemic spread to neighboring communities? What was the time interval between?
  5. Were young and old animals similarly affected, or just one group? Had similar outbreaks occurred before?
  6. Was the outbreak related to the introduction of new animals to the herd or flock or to another observable event? What event? How long after the event did the outbreak occur?
  7. At what time of year did the breakout occur (seasonal)?
  8. What percent of animals died?
  9. Were any human illnesses associated with the outbreak and what were the symptoms?
  10. What measures, if any, did the owners take to control the outbreak? What was done with sick or dead animals?
  11. Do herds or flocks reproduce normally or is early abortion common?
  12. Are animals healthy looking or in generally poor condition?
  13. What symptoms did sick animals show?
  14. Are local veterinarians available for animal treatment and ante- and postmortem inspections of meats? What is their training?
  15. If lay treatment of animals is accomplished, what are common treatments and practices?
  16. What is previous vaccination history, if any?
  17. Are the following diseases endemic to animals in the area:
     a. Anthrax?
     b. Brucellosis?
     c. Tularemia?
     d. Rabies?
     e. Trichinosis?
     f. Worms?
     g. Bovine Tuberculosis?
     h. Q-fever?
     i. Others?

VIII. Local animals (fauna):
  A. Record species of birds, large and small mammals, reptiles, and arthropods present in the area. If names are unknown, describe the species.
  B. Note relationship of these species, including burrows and nesting sites, to human habitation, food supplies, and watering sites.
  C. Note the occurrence of dead or dying animals, especially if the die-off involves a large number of a given species. Note relationship of die-off to occurrence of human disease.
  D. Report any methods used by indigenous personnel to defend against ectoparasites. How effective are standard U.S. protective measures?

(CLASSIFICATION)

Figure C-3. Preventative medicine area assessment (continued)

Area Assessment

(CLASSIFICATION)

IX. **Poisonous plants (flora).** Record those species known to be toxic to man through contact with the skin, inhalation of smoke from burning vegetation, or through ingestion.

X. **Arthropods of medical importance.** Include species and prevalence of the following:
    A. Mosquitoes.
    B. Flies.
    C. Fleas.
    D. Mites.
    E. Ticks.
    F. Lice.
    G. Spiders and other arachnids.

(CLASSIFICATION)

Figure C-3. Preventative medicine area assessment (continued)

This page intentionally left blank.

Glossary

SECTION I – ACRONYMS AND ABBREVIATIONS

18B	Special Forces weapons sergeant
18C	Special Forces engineer sergeant
18D	Special Foces medical sergeant
18E	Special Forces communications sergeant
AAR	after-action review
ACA	airspace coordination area
ACM	airspace coordinating measure
ACO	airspace control order
ACSIM	Assistant Chief of Staff for Installation Management
AFCAP	Air Force Civil Augmentation Program
ALE	Army special operations forces liaison element
ALT	altitude
AMC	Army Materiel Command
AO	area of operations
AOA	amphibious objective area
AOB	advanced operating base
AOR	area of responsibility
AP	antipersonnel
ARNG	Army National Guard
ARNGUS	Army National Guard of the United States
ARSOF	Army special operations forces
ASAALT	Assistant Secretary of the Army for Acquisition, Logistics, and Technology
ASCC	Army Service component command
ASOC	air support operations center
ASP	ammunition supply point
AT	antitank
ATC	air traffic control
ATEP	Antiterrorism Enterprise Portal
ATO	air tasking order
AUTODIN	Automatic Digital Network
BDA	battle damage assessment
BMNT	begin morning nautical twilight
BP	battle position
BPA	blanket purchase agreement
C2	command and control
CA	Civil Affairs

Glossary

CALL	Center for Army Lessons Learned
CAP	crisis action planning
CARVER	criticality, accessibility, recuperability, vulnerability, effect, and recognizability
CAS	close air support
CASCOM	United States Combined Arms Support Command
CASEVAC	casualty evacuation
CBRN	chemical, biological, radiological, and nuclear
CBRNE	chemical, biological, radiological, nuclear, and high-yield explosives
CbT	combating terrorism
CCA	close combat attack
CCDR	combatant commander
CCIF	Combatant Commander's Initiative Fund
CCIR	commander's critical information requirement
CD	counterdrug
CDE	collateral damage estimate
CE	communications-electronics
CERP	commander's emergency response program
CFL	coordinated fire line
CI	counterintelligence
CIDG	civilian irregular defense group
CJCS	Chairman of the Joint Chiefs of Staff
CJCSI	Chairman of the Joint Chiefs of Staff instruction
CJSOTF	combined joint special operations task force
CJTF	commander, joint task force
CNM	critical nodes matrix
COA	course of action
COIN	counterinsurgency
COMSEC	communications security
COMSOC	Commander, Special Operations Command
CONEX	container express
CONPLAN	concept plan
CONUS	continental United States
COR	contracting officer representative
COTR	contracting officer technical representative
COTS	commercial off-the-shelf
CP	contact point
CPAF	cost-plus award fee
CPFF	cost-plus fixed fee
CTA	common table of allowance
CTRIF	Combating Terrorism Readiness Initiative Fund

Glossary

CVAMP	Core Vulnerability Assessment Management Program
DA	Department of the Army
DA PAM	Department of the Army pamphlet
DCG	deputy commanding general
DCSLOG	deputy chief of staff for logistics
DCSOPS	deputy chief of staff for operations
DD Form	Department of Defense form
DEA	Drug Enforcement Administration
DFAC	dining facility
DFARS	Defense Federal Acquisition Regulation Supplement
DOD	Department of Defense
DODI	Department of Defense instruction
DOS	Department of State
DS	direct support
DTG	date-time group
DZ	drop zone
ECP	egress control point
EENT	end evening nautical twilight
EFF	effective
ELRB	executive logistics review board
EO	electro-optical
EPA	evasion plan of action
ERP	en route point
EST	establish
EW	electronic warfare
FAR	Federal Acquisition Regulation
FDC	fire direction center
FFA	free-fire area
FFP	firm, fixed price
FHP	force health protection
FID	foreign internal defense
FLOT	forward line of own troops
FM	field manual
FNS	foreign nation support
FOO	field ordering officer
FPF	final protective fire
FRAGORD	fragmentary order
FS	fire support
FSC	fire support coordinator
FSCC	fire support coordination center
FSCL	fire support coordination line

Glossary

FSCM	fire support coordination measure
FSE	fire support element
FUSB	facilities utilization and support board
FY	fiscal year
GCC	geographic combatant commander
GP	general purpose
GPC	government purchase card
GPS	global positioning system
GS	general support
GSA	General Services Administration
GSB	group support battalion
GTA	graphic training aid
GWOT	Global War on Terror
HA	holding area
HCA	humanitarian and civic assistance
HE	high explosive
HIMARS	high-mobility artillery rocket system
HLZ	helicopter landing zone
HMMWV	high-mobility multipurpose wheeled vehicle
HN	host nation
HP	holding point
HQ	headquarters
HQDA	Headquarters, Department of the Army
HR	human resources
HUMINT	human intelligence
IAW	in accordance with
IBE	installed building equipment
ID	identification
IED	improvised explosive device
IGCE	independent government cost estimate
IMPAC	international merchant purchase authorization card
IO	information operations
IP	initial point
IPBO	installation property book officer
IR	infrared
ISOFAC	isolation facility
ISR	intelligence, surveillance, and reconnaissance
J-4	logistics staff section
J-6	command, control, communications, and computer systems staff section
JAG	Judge Advocate General
JARB	joint acquisition review board

Glossary

JAT Guide	Joint Antiterrorism Program Manager's Guide
JCET	joint combined exchange training
JFC	joint force commander
JFE	joint fires element
JFOB	joint forward operating base
JSOTF	joint special operations task force
JTAC	joint terminal attack controller
JTF	joint task force
KD	known distance
KO	contracting officer
LCMR	lightweight countermortar radar
LEA	law enforcement agency
LOC	line of communications
LOGCAP	logistics civil augmentation program
LP	listening post
LRRP	long-range reconnaissance patrol
LZ	landing zone
Max	maximum
MDMP	military decision-making process
MDSC	medical deployment support command
MEDCAP	medical civic action program
MEDINT	medical intelligence
MEDLOG	medical logistics
METT-TC	mission, enemy, terrain and weather, troops and support available, time available, and civil considerations
MI	military intelligence
MILCON	military construction
Min	minimum
MIPR	military interdepartmental purchase request
MOS	military occupational specialty
MP	military police
MRE	meal, ready to eat
MSS	mission support site
MTOE	modified table of organization and equipment
MTP	mission training plan
MTT	mobile training team
MWR	morale, welfare, and recreation
NATO	North Atlantic Treaty Organization
NCO	noncommissioned officer
NFA	no-fire area
NGO	nongovernmental organization

nm	nautical mile
NSFS	naval surface fire support
NSN	National Stock Number
NTV	nontactical vehicle
NVD	night vision device
NVG	night vision goggle
O&M	operations and maintenance
OAKOC	observation and fields of fire, avenues of approach, key terrain, obstacles, and cover and concealment
OB	order of battle
OGA	other government agency
OIF	Operation IRAQI FREEDOM
OMA	Office of Military Affairs
OP	observation post
OPA	other procurement—Army
OPCEN	operations center
OPCON	operational control
OPLAN	operation plan
OPLOG	operational logistics
OPSEC	operations security
PA	paying agent
PACE	primary, alternate, contingency, and emergency
PARC	principal assistant responsible for contracting
PGM	precision-guided munition
PID	positive identification
PIR	priority intelligence requirement
PLL	prescribed load list
PMCS	preventative maintenance checks and services
POC	point of contact
POI	point of impact
POL	petroleum, oils, and lubricants
POO	point of origin
PP	penetration point
PPBE	Planning, Programming, Budgeting, and Execution
PR&C	purchase request and commitment
PSYOP	Psychological Operations
PWRMS	pre-positioned war reserve materiel stock
PX	post exchange
QRF	quick reaction force
RAM	rockets, artillery, and mortars
RFA	restrictive fire area

RFL	restrictive fire line
RIP	relief in place
ROE	rules of engagement
ROWPU	reverse osmosis water purifcation unit
RP	release point; rendezvous point
RPG	rocket-propelled grenade
S-2	intelligence staff section
S-3	operations staff section
S-4	logistics staff section
S-5	plans staff section
SA	situational awareness
SACC	supporting arms coordination center
SEAD	suppression of enemy air defenses
SF	Special Forces; standard form
SFAUC	Special Forces advanced urban combat
SFG(A)	Special Forces group (airborne)
SFOD	Special Forces operational detachment
SFODA	Special Forces operational detachment A
SFODB	Special Forces operational detachment B
SFODC	Special Forces operational detachment C
SIGCEN	signals center
SIPRNET	SECRET Internet Protocol Router Network
SJA	staff judge advocate
SME	subject-matter expert
SO	special operations
SOC	special operations command
SODARS	special operations debrief and retrieval system
SOF	special operations forces
SOFA	status-of-forces agreement
SOI	signal operating instruction
SOP	standing operating procedure
SOR	statement of requirement
SOSCOM	special operations support command
SOTF	special operations task force
SOTSE	special operations theater support element
SOW	statement of work
SPINS	special instructions
SPTCEN	support center
SSN	standard study number
SSR	service, supply, and repair
STP	Soldier training publication

Glossary

SWEAT-MS	sewage, water, electricity, academics, trash, medical, and security
SWEAT-MSS	security, water, electricity, administration, trash, medical, sewage, and shelter
T&M	time and materials
TACFAC	tactical facility
TAMCA	theater Army movement control agency
TAMMC	theater Army material management command
TDY	temporary duty
TMO	transportation management office
TOA	transfer of authority
TOC	tactical operations center
TPFDL	time-phased force and deployment list
TRP	target reference point
TSC	theater sustainment command
TSOC	theater special operations command
TTP	tactics, techniques, and procedures
UAS	unmanned aircraft system
UBL	unit basic load
UFR	unfunded requirement
UN	United Nations
U.S.	United States
USACE	United States Army Corps of Engineers
USAF	United States Air Force
USAJFKSWCS	United States Army John F. Kennedy Special Warfare Center and School
USAR	United States Army Reserve
USASOC	United States Army Special Operations Command
USC	United States Code
USCENTCOM	United States Central Command
USG	United States Government
USMC	United States Marine Corps
USN	United States Navy
UTAMS	unattended transient acoustic measurement and signature intelligence system
UXO	unexploded ordnance
VBIED	vehicleborne improvised explosive device
WP	white phosphorus

SECTION II – TERMS

air support operations center
The principal air control agency of the theater air control system responsible for the direction and control of air operations directly supporting the ground combat element. It processes and coordinates requests for immediate air support and coordinates air missions requiring integration with other

supporting arms and ground forces. It normally collocates with the Army tactical headquarters senior fire support coordination center within the ground combat element. Also called **ASOC**. (JP 1-02)

air tasking order
A method used to task and disseminate to components, subordinate units, and command and control agencies projected sorties, capabilities and/or forces to targets and specific missions. Normally provides specific instructions to include call signs, targets, controlling agencies, etc., as well as general instructions. Also called **ATO**. (JP 1-02)

casualty evacuation
The unregulated movement of casualties that can include movement both to and between medical treatment facilities. Also called **CASEVAC**. (JP 1-02)

close air support
Air action by fixed- and rotary-wing aircraft against hostile targets that are in close proximity to friendly forces and that require detailed integration of each air mission with the fire and movement of those forces. Also called **CAS**. (JP 1-02)

combating terrorism
Actions, including antiterrorism (defensive measures taken to reduce vulnerability to terrorist acts) and counterterrorism (offensive measures taken to prevent, deter, and respond to terrorism), taken to oppose terrorism throughout the entire threat spectrum. Also called **CbT**. (JP 1-02)

combined joint special operations task force
A task force composed of special operations units from one or more foreign countries and more than one U.S. Military Department formed to carry out a specific special operation or prosecute special operations in support of a theater campaign or other operations. The combined joint special operations task force may have conventional nonspecial operations units assigned or attached to support the conduct of specific missions. Also called **CJSOTF**. (JP 1-02)

contact point
1. In land warfare, a point on the terrain, easily identifiable, where two or more units are required to make contact. 2. In air operations, the position at which a mission leader makes radio contact with an air control agency. 3. (DOD only) In personnel recovery, a location where isolated personnel can establish contact with recovery forces. Also called **CP**. (JP 1-02)

contracting officer
A U.S. military officer, enlisted member, or civilian employee who has a valid appointment as a contracting officer under the provisions of the Federal Acquisition Regulation. The individual has the authority to enter into and administer contracts and determinations as well as findings about such contracts. (JP 1-02)

coordinated fire line
A line beyond which conventional and indirect surface fire support means may fire at any time within the boundaries of the establishing headquarters without additional coordination. The purpose of the coordinated fire line is to expedite the surface-to-surface attack of targets beyond the coordinated fire line without coordination with the ground commander in whose area the targets are located. Also called **CFL**. (JP 1-02)

cost-plus a fixed fee contract
A cost-reimbursement type contract that provides for the payment of a fixed fee to the contractor. The fixed fee, once negotiated, does not vary with actual cost but may be adjusted as a result of any subsequent changes in the scope of work or services to be performed under the contract. (JP 1-02)

evasion plan of action
A course of action, developed prior to executing a combat mission, that is intended to improve a potential isolated person's chances of successful evasion and recovery by providing the recovery

Glossary

forces with an additional source of information that can increase the predictability of the evader's action and movement. Also called **EPA**. (JP 1-02)

final protective fire
An immediately available prearranged barrier of fire designed to impede enemy movement across defensive lines or areas. (JP 1-02)

fire direction center
That element of a command post, consisting of gunnery and communications personnel and equipment, by means of which the commander exercises fire direction and/or fire control. The fire direction center receives target intelligence and requests for fire, and translates them into appropriate fire direction. The fire direction center provides timely and effective tactical and technical fire control in support of current operations. Also called **FDC**. (JP 1-02)

fire support
Fires that directly support land, maritime, amphibious, and special operations forces to engage enemy forces, combat formations, and facilities in pursuit of tactical and operational objectives. (JP 1-02)

fire support coordination center
A single location in which are centralized communications facilities and personnel incident to the coordination of all forms of fire support. Also called **FSCC**. (JP 1-02)

fire support coordination line
A fire support coordination measure that is established and adjusted by appropriate land or amphibious force commanders within their boundaries in consultation with superior, subordinate, supporting, and affected commanders. Fire support coordination lines facilitate the expeditious attack of surface targets of opportunity beyond the coordinating measure. A fire support coordination line does not divide an area of operations by defining a boundary between close and deep operations or a zone for close air support. The fire support coordination line applies to all fires of air, land, and sea-based weapon systems using any type of ammunition. Forces attacking targets beyond a fire support coordination line must inform all affected commanders in sufficient time to allow necessary reaction to avoid fratricide. Supporting elements attacking targets beyond the fire support coordination line must ensure that the attack will not produce adverse effects on, or to the rear of, the line. Short of a fire support coordination line, all air-to-ground and surface-to-surface attack operations are controlled by the appropriate land or amphibious force commander. The fire support coordination line should follow well-defined terrain features. Coordination of attacks beyond the fire support coordination line is especially critical to commanders of air, land, and special operations forces. In exceptional circumstances, the inability to conduct this coordination will not preclude the attack of targets beyond the fire support coordination line. However, failure to do so may increase the risk of fratricide and could waste limited resources. Also called **FSCL**. (JP 1-02)

fire support element
That portion of the force tactical operations center at every echelon above company or troop (to corps) that is responsible for targeting coordination and for integrating fires delivered on surface targets by fire support means under the control, or in support, of the force. Also called **FSE**. (JP 1-02)

force health protection
Measures to promote, improve, or conserve the mental and physical well-being of Service members. These measures enable a healthy and fit force, prevent injury and illness, and protect the force from health hazards. Also called **FHP**. (JP 1-02)

free-fire area
A specific area into which any weapon system may fire without additional coordination with the establishing headquarters. Also called **FFA**. (JP 1-02)

humanitarian and civic assistance
Assistance to the local populace provided by predominantly U.S. forces in conjunction with military operations and exercises. This assistance is specifically authorized by Title 10, United States Code,

Glossary

Section 401 and funded under separate authorities. Assistance provided under these provisions is limited to (1) medical, dental, and veterinary care provided in rural areas of a country; (2) construction of rudimentary surface transportation systems; (3) well drilling and construction of basic sanitation facilities; and (4) rudimentary construction and repair of public facilities. Assistance must fulfill unit training requirements that incidentally create humanitarian benefit to the local populace. Also called **HCA**. (JP 1-02)

intelligence, surveillance, and reconnaissance
An activity that synchronizes and integrates the planning and operation of sensors, assets, and processing, exploitation, and dissemination systems in direct support of current and future operations. This is an integrated intelligence and operations function. Also called **ISR**. (JP 1-02)

joint combined exchange training
A program conducted overseas to fulfill U.S. forces training requirements and at the same time exchange the sharing of skills between U.S. forces and host nation counterparts. Training activities are designed to improve U.S. and host nation capabilities. Also called **JCET**. (JP 1-02)

joint fires element
An optional staff element that provides recommendations to the operations directorate to accomplish fires planning and synchronization. Also called **JFE**. (JP 1-02)

joint special operations task force
A joint task force composed of special operations units from more than one Service, formed to carry out a specific special operation or prosecute special operations in support of a theater campaign or other operations. The joint special operations task force may have conventional non-special operations units assigned or attached to support the conduct of specific missions. Also called **JSOTF**. (JP 1-02)

joint terminal attack controller
A qualified (certified) Service member who, from a forward position, directs the action of combat aircraft engaged in close air support and other offensive air operations. A qualified and current joint terminal attack controller will be recognized across the Department of Defense as capable and authorized to perform terminal attack control. Also called **JTAC**. (JP 1-02)

mobile training team
A team consisting of one or more U.S. military or civilian personnel sent on temporary duty, often to a foreign nation, to give instruction. The mission of the team is to train indigenous personnel to operate, maintain, and employ weapons and support systems, or to develop a self-training capability in a particular skill. The Secretary of Defense may direct a team to train either military or civilian indigenous personnel, depending upon host-nation requests. Also called **MTT**. (JP 1-02)

no-fire area
An area designated by the appropriate commander into which fires or their effects are prohibited. Also called **NFA**. (JP 1-02)

National Stock Number
The 13-digit stock number replacing the 11-digit Federal Stock Number. It consists of the 4-digit Federal Supply Classification code and the 9-digit National Item Identification Number. The National Item Identification Number consists of a 2-digit National Codification Bureau number designating the central cataloging office (whether North Atlantic Treaty Organization or other friendly country) that assigned the number and a 7-digit (xxx-xxxx) nonsignificant number. The number shall be arranged as follows: 9999-00-999-9999. Also called **NSN**. (JP 1-02)

naval surface fire support
Fire provided by Navy surface gun and missile systems in support of a unit or units. Also called **NSFS**. (JP 1-02)

Glossary

nongovernmental organization
A private, self-governing, not-for-profit organization dedicated to alleviating human suffering; and/or promoting education, health care, economic development, environmental protection, human rights, and conflict resolution; and/or encouraging the establishment of democratic institutions and civil society. Also called **NGO**. (JP 1-02)

order of battle
The identification, strength, command structure, and disposition of the personnel, units, and equipment of any military force. Also called **OB**. (JP 1-02)

other government agency
Within the context of interagency coordination, a non Department of Defense agency of the United States Government. Also called **OGA**. (JP 1-02)

relief in place
An operation in which, by direction of higher authority, all or part of a unit is replaced in an area by the incoming unit. The responsibilities of the replaced elements for the mission and the assigned zone of operations are transferred to the incoming unit. The incoming unit continues the operation as ordered. (JP 1-02)

restrictive fire area
An area in which specific restrictions are imposed and into which fires that exceed those restrictions will not be delivered without coordination with the establishing headquarters. Also called **RFA**. (JP 1-02)

restrictive fire line
A line established between converging friendly surface forces that prohibits fires or their effects across that line. Also called **RFL**. (JP 1-02)

supporting arms coordination center
A single location on board an amphibious command ship in which all communication facilities incident to the coordination of fire support of the artillery, air, and naval gunfire are centralized. This is the naval counterpart to the fire support coordination center utilized by the landing force. Also called **SACC**. (JP 1-02)

suppression of enemy air defenses
Activity that neutralizes, destroys, or temporarily degrades surface-based enemy air defenses by destructive and/or disruptive means. Also called **SEAD**. (JP 1-02)

tactical facility
Any secure urban or rural facility that enables Army special operations forces to extend command and control, provides support for operations, and allows operational elements to influence a specified area. Special Forces tactical facilities include a variety of secure locations for Special Forces operations, including (but not limited to) firebases, camps, and team houses. Also called **TACFAC**.

time-phased force and deployment list
Appendix 1 to Annex A of the operation plan. It identifies types and/or actual units required to support the operation plan and indicates origin and ports of debarkation or ocean area. It may also be generated as a computer listing from the time-phased force and deployment data. Also called **TPFDL**. (JP 1-02)

References

Army Publications

DA Form 1355-1 (Hasty Protective Row Minefield Record)
DA Form 3953 (Purchase Request and Commitment [PR&C])
DA Form 5517-R (Standard Range Card)

Note: DA forms are available on the APD Web site (www.apd.army.mil).

DA PAM 420-11, *Facilities Engineering Project Definition & Work Classification*, 7 October 1994.
FM 3-0, *Operations*, 27 February 2008.
FM 3-07, *Stability Operations*, 6 October 2008.
FM 3-09.32, *(JFIRE) Multiservice Tactics, Techniques, and Procedures for the Joint Application of Firepower*, 20 December 2007.
FM 3-19.1, *Military Police Operations*, 22 March 2001.
FM 5-0, *Army Planning and Orders Production*, 20 January 2005.
FM 5-34, *Engineer Field Data*, 19 July 2005.
GTA 31-01-003, *Detachment Mission Planning Guide*, 1 March 2006.
GTA 41-01-005, *Religious Factors Analysis*, 2 January 2008.
STP 31-18C34-SM-TG, *Soldier's Manual and Trainer's Guide, MOS 18C, Special Forces Engineer Sergeant Skill Levels 3 and 4*, 8 July 2003.

Joint Publications

CJCSI 7401.01D, *Combatant Commander Initiative Fund*, 30 September 2008.
JP 1-02, *Department of Defense Dictionary of Military and Associated Terms*, 12 April 2001.
JP 3-09.3, *Joint Tactics, Techniques, and Procedures for Close Air Support (CAS)*, 3 September 2003.

Other Publications

DD Form 1155 (Order for Supplies or Services)
DD Form 1391 (Military Construction Project Data)
DD Form 1972 (Joint Tactical Air Strike Request)

Note: DD forms are available on the OSD Web site (www.dtc.mil/whs/directives/infomgt/forms/formsprogram.htm).

DODI 2000.16, *DoD Antiterrorism (AT) Standards*, 2 October 2006.
SF 33 (Solicitation, Offer, and Award)
SF 44 (U.S. Government Purchase Order Invoice Voucher)

Note: The images depicted in Figure 2-8 (page 2-11), Figure 2-9 (page 2-15), and Figure 3-1 (page 3-3) are copyrighted. They are used in this publication with permission of the copyright holder, Osprey Publishing.

This page intentionally left blank.

Index

A
access road, 1-8, 1-10, 2-3, 2-12, 2-22, 2-24
administration area, 2-3
administration sector, 1-1, 1-3, 1-4, 1-8, 1-10, 2-3, 2-7, 2-9, 2-19, 2-22, 2-24, 5-8
airfield, 2-19, 2-21, 2-24, 2-25, 3-7, 3-8
airspace coordination area, 4-19, 4-20
ammunition bunkers, 2-10, 2-14, 2-15, 3-11
ammunition crates, 2-10, 3-10
area assessment, 2-1, 2-2, C-1 through C-9
  initial, C-1
  preventative medicine, C-4
  principal, C-1
area study, 2-1, 7-3, A-1 through A-5, C-1
automatic resupply, 5-7

B
Brick and masonry, 3-9

C
cache, 5-8
Center for Army Lessons Learned, 2-2
cinder block, 3-11
Cinva-Ram, 2-7, 3-12
classes of supply, 5-2, 5-8
claymore mines, 1-10, 2-13
close air support, 4-1, 4-3 through 4-5, 4-10 through 4-17
close combat attack, 4-3 through 4-5, 4-10 through 4-17
closeout, 7-6, 7-7
Combatant Commander's Initiative Fund, 6-10
combating terrorism, 6-9
commander's emergency response program, 6-5, 6-6
concept plan, 5-8
concrete, 3-1, 3-6 through 3-8, 6-5

CONEX, 2-10, 2-15, 3-11
construction
  fighting trenches, 3-5
  floors, 3-1
  machine-gun bunkers, 3-2
  mortar positions, 3-2
  observation towers, 3-4
  roof, 3-2
  walls, 3-1
contact point, 4-16
contracting officer technical representative, 6-11
coordinated fire line, 4-18
crisis action planning, 5-8
critical nodes matrix, 1-3, 1-4

D
dining facility, 1-8, 2-14, 2-18, 5-5, 5-9, 5-10, 5-13, 6-5, B-3
dispensary, 1-5, 1-12, 2-9 through 2-21

E
egress control point, 4-16, 4-17
emergency resupply, 5-7
emergency signal, 2-12, 3-12
en route point, 4-16
evasion plan of action, 7-3, 7-6, B-1

F
fencing, 2-6, 2-7, 2-12, 2-14, 3-8, 3-12
field ordering officer, 6-1, 6-2, 6-11, 6-13
field services, 5-3
fighting bunker, 2-14
fire arrow, 2-12, 3-12
fire support coordination line, 4-18, 4-19
fire support coordination measure, 4-10, 4-16, 4-18
force health protection, 5-1, 5-6
free-fire area, 4-19
fuel point, 2-14, 2-18

G
generator bunkers, 2-9, 2-10
government building, 2-19, 2-21

government purchase card, 6-15
group support battalion, 5-4 through 5-6, 5-8

H
helicopter landing zone, 1-5, 1-12, 2-19, 2-21, 3-8
holding area, 4-17
hostile environment, 2-3, 4-2
human resource, 5-1, 5-6
humanitarian and civic assistance, 6-10

I
independent government cost estimate, 6-1 through 6-4
initial phase, 1-4, 1-12
initial point, 4-11, 4-16
inner barrier, 2-3
inner perimeter, 1-10, 2-3, 2-6, 2-7, 2-10 through 2-12, 2-14, 2-17, 2-18
installation property book officer, 6-2
international merchant purchase authorization card, 6-14
ISOFAC, 2-7
isolation facility, 5-13

J
joint acquisition review board, 6-1 through 6-3, 6-11
joint combined exchange training, 5-5
joint fires, 4-3
joint special operations area, 5-7, 5-8
joint special operations task force, 5-1, 5-2, 5-5, 5-6, 7-6
joint terminal air controller, 4-5, 4-10 through 4-14, 4-16, 4-17

L
latrine, iii, 1-7, 1-8, 2-11, 3-10
line of communications, 2-14

M

main gate, 2-19, 2-22 through 2-24
main road, 2-24, 2-25
main-gate bunkers, 2-23
maintenance, 1-10, 1-13, 2-1, 2-9, 2-10, 2-12, 2-14, 2-18, 4-1, 5-1 through 5-6, 5-9, 5-11 through 5-13, 5-15, 6-5 through 6-7, 6-9
medical bunker, 2-9
military construction, 6-6 through 6-9
military decision-making process, 2-1, 2-2, 7-6
military interdepartmental purchase request, 6-4
mobile training team, 5-5
mortar positions, 2-10
motor pool, 2-14, 2-18

N

no-fire area, 4-19

O

observation towers, 2-16, 3-4, 3-5
Office of Military Affairs, 6-6
on-call resupply, 5-7
operation plan, 5-1, 5-8
operations center, 1-8, 2-1, 2-3, 2-7, 5-13
other procurement—Army, 6-6, 6-7
outer barrier, 1-10, 2-3
outer perimeter, 2-3, 2-14, 2-17

P

pallets, 2-10, 2-15, 3-1, 3-8, 3-10, 3-11

paying agent, 6-1, 6-2, 6-11, 6-13
perimeter road, 2-12, 2-24
permanent phase, 1-10, 1-13
permissive environment, 4-2
petroleum, oils, and lubricants, 5-7, 5-8, 5-10, 5-11, 6-12, A-5, C-3
Purchase Request and Commitment, 6-1 through 6-4, 6-13

R

radar, 1-4, 4-6, 4-7 through 4-9
restrictive fire area, 4-19
restrictive fire line, 4-19
rock, 3-8, 3-9, 3-11, A-2
rules of engagement, 2-13, 4-2, 4-10, 4-13, 4-16
rural tactical facilities, v, 1-2 through 1-4, 1-6 through 1-8, 1-10, 1-12, 1-13, 2-3, 2-6, 2-16, 2-25, 3-1, 3-7, 3-12, 7-5, 7-6, A-2, A-4, C-6

S

sandbags, 2-7, 2-9, 2-10, 2-15, 2-19, 2-21, 3-1, 3-2, 3-5, 3-8, 3-9, 3-10, 3-11, 5-8
signals center, 2-1, 5-13
site survey, 2-1, 2-2, B-1
special operations debrief and retrieval system, 2-1, 2-2
special operations theater support element, 5-8
stackable barrier system, 1-5, 1-10
staff judge advocate, 6-4, 6-6, 6-8, 6-9, 6-11, 6-13, 6-15
statement of requirement, 2-1, 5-1, 5-6, 5-8

statement of work, 6-1 through 6-4, 6-7, 6-12
steel drums, 3-10, 3-11
support center, 2-1, 5-1, 5-13
suppression of enemy air defenses, 4-14
surrounding area, 1-8, 1-10, 2-3, 2-19, 2-21, 2-22, 2-24, 2-25

T

tactical operations center, 2-3, 2-17
tanglefoot, 1-10, 2-12
temporary phase, 1-8, 1-13
threat level, 4-2
Title 10, United States Code, 5-8, 6-5
transfer of authority, 7-1
trash point, 2-12, 2-18

U

uncertain environment, 4-2
undeveloped theater, 5-5, 5-6
unfunded requirement, 6-9
unmanned aircraft system, 4-9, 4-10
urban tactical facilities, 1-12, 7-4, 7-6

V

vehicle revetments, 2-11
vehicleborne improvised explosive device, 7-5

W

water point, 2-18
weapons range, 2-18
wood, 3-1, 3-6, 3-7, 3-9

FM 3-05.230
8 February 2009

By Order of the Secretary of the Army:

                                          GEORGE W. CASEY, JR.
                                       General, United States Army
                                               Chief of Staff

Official:

[signature]

JOYCE E. MORROW
*Administrative Assistant to the
Secretary of the Army*
      0900905

DISTRIBUTION:
Active Army, Army National Guard, and U.S. Army Reserve: To be distributed in accordance with the initial distribution number (IDN) 115905, requirements for FM 3-05.230.

PIN: 080972-000

www.ingramcontent.com/pod-product-compliance
Lightning Source LLC
Chambersburg PA
CBHW050101230526
45470CB00004B/1624